SERVICE ROBOTS

Edited by **Antonio J. R. Neves**

Service Robots
http://dx.doi.org/10.5772/65536
Edited by Antonio J. R. Neves

Contributors

Daniel P Bernardon, Carlos Barriquello, Luciane Canha, Flavio Silva, Maicon Ramos, Daniel Porto, Juan Jesús Roldán, Antonio Barrientos Cruz, Jaime Del Cerro, Mario Garzón, Jorge De León, Pablo Garcia-Aunon, David Garzón, Florin Daniel Anton, Silvia Anton, Mihai Constantinescu, Marcel Horák, František Novotný, Michal Starý, Eduard Fosch Villaronga, Jordi Albo-Canals, Inês Domingues, Ricardo Silva, Luís Matos, Daniel Campos, Joana Santos, João Leão, José Xavier, Manuel Camarneiro, Marcelo Penas, Maria Miranda, Ricardo Morais, Tiago Esteves, Markus Vincze, Markus Bajones, Markus Suchi, Daniel Wolf, Lara Lammer, Astrid Weiss, David Fischinger, Tianyi Wang, Hieyong Jeong, Yuko Ohno

Published by InTech
Janeza Trdine 9, 51000 Rijeka, Croatia

Publishing Process Manager Nina Kalinic
Technical Editor SPi Global
Cover InTech Design team

First published December, 2017
Printed in Croatia
Legal deposit, Croatia: National and University Library in Zagreb

Additional hard copies can be obtained from orders@intechopen.com

Service Robots, Edited by Antonio J. R. Neves
p. cm.
Print ISBN 978-953-51-3722-1
Online ISBN 978-953-51-3723-8

For EU product safety concerns:
IN TECH d.o.o., Prolaz Marije Krucifikse Kozulić 3, 51000 Rijeka, Croatia,
info@intechopen.com or visit our website at intechopen.com.

Meet the editor

Prof. António J. R. Neves received his PhD degree in Electrical Engineering from the University of Aveiro, in 2007. Currently, he is an assistant professor at the Department of Electronics, Telecommunications and Informatics of the University of Aveiro. He is an IEEE senior member and a member of several other research organizations worldwide, namely, IEEE RAS, EURASIP, IASTED, INSTICC, and Portuguese Robotics Society, among others. His main research interests are robotics, computer vision, and image and video processing. He participated in or coordinated several research projects and published more than 140 publications, including 2 books, book chapters, journal articles, and conference papers. He also has a vast experience in supervising PhD and MSc degree students and is a reviewer of several journals and conferences.

Contents

Preface

According to some authors, service robots are considered as reprogrammable, sensor-based mechatronic devices that perform useful service to human activities in an everyday environment. Moreover, service robots perform tasks in a specific environment and should be able to perform services semi- or fully automatically. The potential applications of this technology in warehouses, hospitals, retail stores, city streets, industrial parks, and college campuses and as personal assistants just to name a few are representatives of this new invading force that is starting to become apparent in our daily lives.

The exponential effort on the development of these robots is confirmed by the amount of money invested in projects and companies, the creation on new start-ups worldwide, and, not less important, the quantity and quality of the manuscripts published in a considerable number of journals and conferences worldwide.

This book is an outcome of research done by several researchers and professionals who have highly contributed to the field of service robots. The main goal of this book is to present recent advances in the field of service robots. I would like to thank all the authors for their excellent contributions and all those people who helped in this project.

This book contains seven chapters divided into four sections. Section 1 consists of three chapters focusing on the use of service robots to improve people's lives. Section 2 presents several other applications of service robots, from agriculture to industry. Section 3 consists of a single chapter related to the importance of communications in robotics. The last chapter contains a chapter dedicated to law issues in this field.

Finally, I hope that all the readers of this book will find it interesting and informative, considering it as a good tool for their research or project.

Prof. Antonio J. R. Neves, PhD
Department of Electronics, Telecommunications and Informatics,
University of Aveiro,
Portugal

Section 1

Improvement of People's Life

Chapter 1

A Personal Robot as an Improvement to the Customers' In-Store Experience

Joana Santos, Daniel Campos, Fábio Duarte,
Filipe Pereira, Inês Domingues, Joana Santos,
João Leão, José Xavier, Luís de Matos,
Manuel Camarneiro, Marcelo Penas, Maria Miranda,
Ricardo Morais, Ricardo Silva and Tiago Esteves

Additional information is available at the end of the chapter

http://dx.doi.org/10.5772/intechopen.70277

Abstract

Robotics is a growing industry with applications in numerous markets, including retail, transportation, manufacturing, and even as personal assistants. Consumers have evolved to expect more from the buying experience, and retailers are looking at technology to keep consumers engaged. In today's highly competitive business climate, being able to attract, serve, and satisfy more customers is a key to success. It is our belief that smart robots will play a significant role in physical retail in the future. One successful example is wGO,[1] a robotic shopping assistant developed by Follow Inspiration. The wGO is an autonomous and self-driven shopping cart, designed to follow people with reduced mobility (the elderly, people in wheelchair, pregnant women, those with temporary reduced mobility, etc.) in commercial environments. With the Retail Robot, the user can control the shopping cart without the need to push it. This brings numerous advantages and a higher level of comfort since the user does not need to worry about carrying the groceries or pushing the shopping cart. The wGO operates under a vision-guided approach based on user-following with no need for any external device. Its integrated architecture of control, navigation, perception, planning, and awareness is designed to enable the robot to successfully perform personal assistance, while the user is shopping.

Keywords: robotics, vision, retail, reduced mobility

[1]The robot is currently patent pending.

1. Introduction

In recent years, a high concern with user satisfaction has been observed in the retail industry. This is particularly accentuated with the rise in online shopping which pushes retailers to provide a better in-person shopping experience to attract customers. Among customers in the public, one of the main groups of interest is people with disabilities. This is visible not only in the marketing strategies but also at the political level, where accessibility for disabled people is becoming the topic of regulation and legislation.

It is estimated that in Portugal about 8–10% of the population has some form of disability [1] and that in Europe alone there are about 50 million people with disabilities and 134 million people with reduced mobility. Apart from people using wheelchairs, there are other cases in which people are temporarily or permanently disabled, and these include: an elderly person using a cane, or someone with a foot or leg injury who requires the use of crutches, pregnant ladies, and parents with prams.

In fact, if we add the disabled, the elderly, pregnant women, and couples with children, we find that between 30 and 40% of all Europeans could benefit from improved accessibility. In addition to those people with reduced mobility due to disability or injury, there are many people without mobility issues who could benefit from assistance in carrying heavy bags.

Shopping environments are highly heterogeneous and give rise to a high frequency of dynamic interactions that trigger various senses and emotions in humans. This often causes a high level of stress in people and those with mobility limitations.

Some of the identified difficulties include [2]:

- People who use wheelchairs
  - no adequate forward reach at basins, counters, and tables; and
  - surfaces that do not provide sufficient traction (e.g., polished surfaces).
- People who have trouble walking
  - no seating in waiting areas, at counters and along lengthy walkways;
  - access hazards associated with doors, including the need to manipulate a handle while using a walking aid; and
  - surface finishes that are not slip-resistant or are unevenly laid.

Besides the difficulties brought by the shopping environment itself, conventional shopping carts, which can carry many products and which are provided with wheels so that the shoppers can push them, also have serious drawbacks, one of them being their considerable size. This is simultaneously an important asset and a significant drawback, as although shopping carts can hold large and bulky products, the increased mass complicates maneuverability and handling. Maneuverability is particularly compromised when making turns in supermarket aisles or when avoiding other carts, shelves, and indeed other shoppers [3]. Smaller baskets appeared on the market to overcome the traditional shopping carts' drawbacks. These baskets were developed to hold a set of items while at the same time being easy to move. They contain wheels or rolling elements incorporated into the bases which allow them to be moved when parallel to the floor or when inclined. However, even though these baskets improve maneuverability due to their reduced size and capacity, they also have drawbacks typical of their morphology, such as the need for the user to bend down for placing or removing items, among others. Furthermore, such baskets can have drawbacks typical of the way they are stored, since stacking them vertically can entail a problem for elderly shoppers or shoppers with any type of physical limitation [3].

With these described difficulties in mind, this paper presents a new robotic concept to help and assist people (giving special emphasis to people with reduced mobility) in these types of environments, through a *user-following* scenario. The wGO (**Figure 1**) is an autonomous and self-driven shopping cart, designed to follow people with reduced mobility (the elderly, people in wheelchairs, pregnant women, temporary reduced mobility, etc.) in commercial environments. With the robot, the user can control the shopping cart without the need to push it. This brings numerous advantages and a higher level of comfort, since the user does not need to worry about carrying the groceries or pushing the shopping cart.

In this chapter, the wGO is introduced for the first time and we present its behavior and main features. Preliminary technical results obtained in real scenarios are also given. Finally, a user satisfaction survey is presented. The structure of the chapter is as follows: the current available solutions and their limitations are discussed in Section 2. In Section 3, the wGO's behavior is described, while its architecture is given in Section 4. Section 5 illustrates the wGO

Figure 1. wGO.

behavior in a real scenario (a supermarket during regular opening hours) and analyses a user study. Conclusions and directions for future work are given in Section 6.

2. Existing solutions

Looking at the commercial market, the most obvious existing solutions are those provided by shopping cart producers.[2] These providers typically have products targeted for customers in wheelchair, but not products for other types of users with reduced mobility (e.g., pregnant women). A different type of solution is the adapted system. Some examples are the "amigo mobility" scooter[3] and the adapted wheelchairs,[4] promoted by Egiro.[5] These products are, however, not particularly user-friendly. The user needs to first move into the mobility auxiliary device and then to learn how to use it (which may be particularly hard for the scooter case). In the case of wheelchair users, the user also needs to leave their own personal chair, which may cause discomfort and unnecessary stress. Another problem with these solutions is that the user is visibly distinguishable from the other supermarket clients, which may discourage some people from using it [4].

While this topic of assisted shopping using robotics has received very little attention in the academic research community, several systems exist where robots are used to help people with reduced mobility.

In [5], an anticipative shared control for robotic wheelchairs, targeted at people with disabilities, is presented. The same idea, of intelligent wheelchairs, is also the focus of the work in [6] where a data analysis system which provides an adapted command language is presented. The work in [7] presents an analysis of the implementation of a system for navigating a wheelchair with automation, based on facial expressions, especially eyes closed using a Haar cascade classifier, aimed at people with locomotor disability of the upper and lower limbs.

A smart companion robot for the elderly people, capable of carrying out surveillance and telepresence tasks, is described in [8]. Also, with the aim of helping elderly people through telepresence, a low-cost platform capable of providing augmented reality for pill dose management was developed in [9]. In [10], an approach based on the Dynamical System Approach for obstacle avoidance of a Smart Walker device to help navigation of elderly people is presented.

Perhaps, the closest application to the focus of this chapter is presented in [11] where a product locator application is proposed. The application runs on heterogeneous personal mobile devices, keeping the user private information safe and locating the desired

[2]For example, wanzl (www.wanzl.com).
[3]www.myamigo.com
[4]For example, meyra (www.meyra.de).
[5]www.egiro.pt

products over each supermarket's map. We believe that such a system could be complementary to the wGO and could be used in combination to further improve customers' shopping experiences.

3. wGO: Specifications and behavior descriptions

The wGO is designed to have an ergonomic shape, friendly both to the target users (people with reduced mobility) and to the environment (commercial retail environment). An illustration of the robot's hardware is shown in **Figure 2**. Its main internal sensors are as follows:

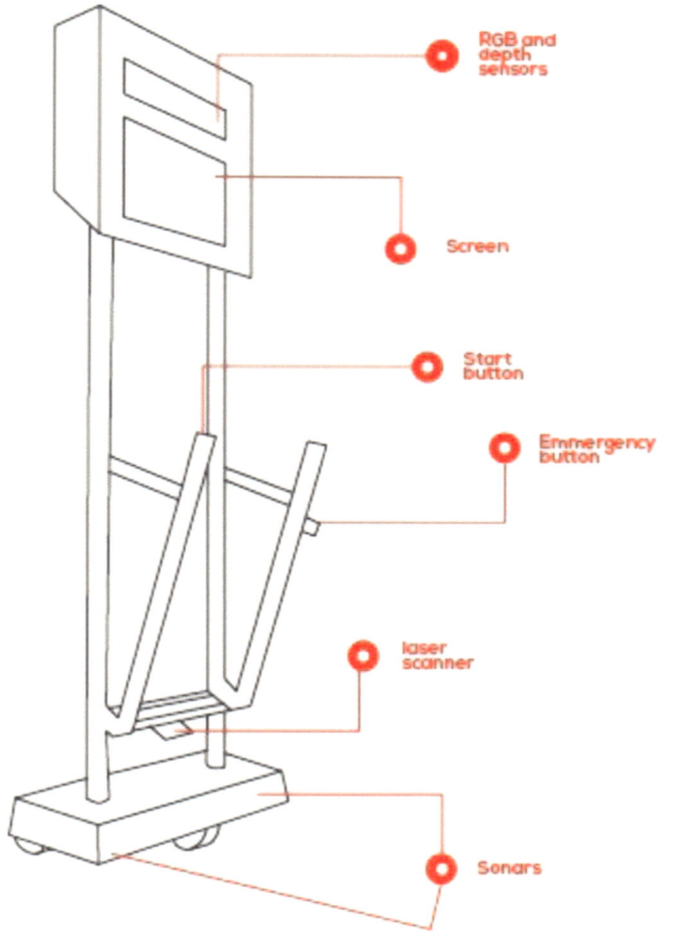

Figure 2. wGO hardware illustration.

Figure 3. wGO initialization: wheelchair typical case.

ultrasound sensors, a Laser Range Finder (LRF), and active vision sensors. This combination was selected due to their complementary features. While the ultrasonic sensor detects any type of material that is not sound absorbing, it has the wide beam-width and echo problems as main drawbacks. LRF provides 270° information and its precision is high. It is, however, sensitive to dust. Active vision provides very rich information (image + 3D), but it has a relatively small field of view and low precision.

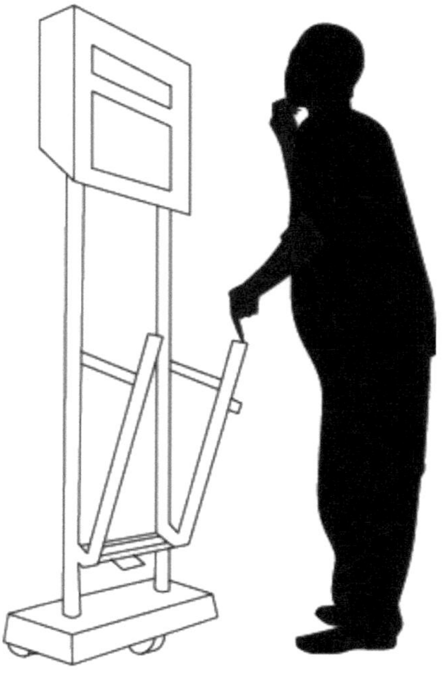

Figure 4. wGO initialization: non-wheelchair typical case.

http://dx.doi.org/10.5772/intechopen.70277

The ultrasound sensors have a minimum and maximum range of 3 cm and 4 m, respectively, and an estimated field of view of 60°. The LRF has a maximum range of 6 m and performs 270° laser scanning. The specifications of the active vision systems are as follows:

- Range: 0.6–8 m (optimal 0.6–5.0 m)
- Color camera: 1280x960 @ 10 FPS
- Depth camera: 640x480 (VGA) 16 bit @ 30 FPS
- Horizontal field of view: 60°
- Vertical field of view: 49.5°

The system is initialized when the user presses the start button. At this moment, the user is typically facing the wGO (**Figures 3** and **4**). After initialization, the user starts shopping and the wGO follows him or her.

In cases where the person goes out of the image sensors' field of view (**Figure 5**), there is a 270° laser scanner that aids the tracking process.

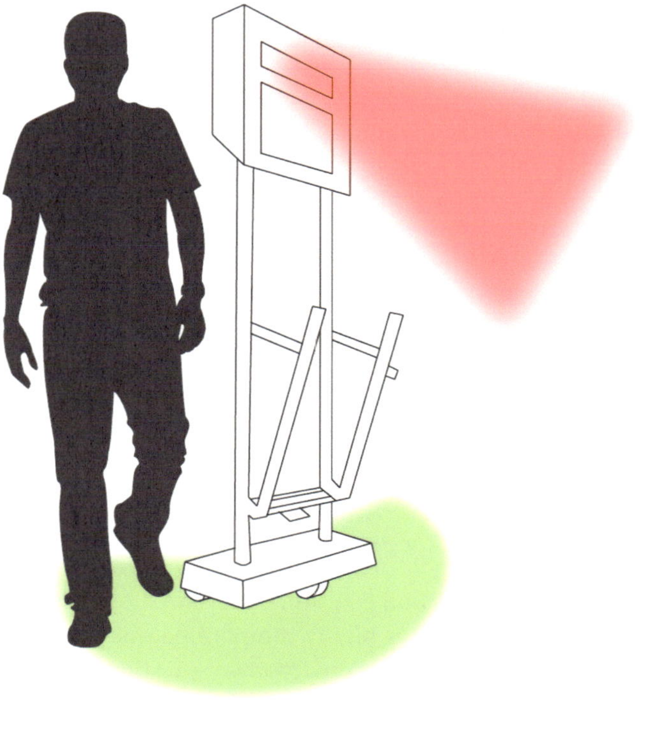

Figure 5. Illustration of a case where the user is out of the field of view of the RGB and depth sensors, but visible by the 270° laser scanner.

4. wGO: System architecture

Figure 6 depicts the functional diagram of the application that is embedded in the robot and is responsible for gathering information from the sensors, for example, the encoders and the RGB and depth cameras, as well as controlling the movement of the robot. Therefore, this figure depicts a high-level representation of how perceptual data can be combined and used to enable a robot to follow a user in a realistic environment.

Internally, the application is divided into several modules: Vision, Sensors, Behavior, Executing, and Control system. The Vision module grabs and processes RGB and depth information. In addition, the same module performs people detection [12], false-positive reduction, and identification tasks. The Sensors module grabs data received from the sensors and verifies the existence of obstacles. The Behavior module includes the tracking [13] of the detected person and the generation of the path [14] for the robot to follow. In path generation, a local localization method based on odometry is used to retrieve the location of the robot and the fusion of the vision and laser tracking results is made [1, 15–17]. The Executing system receives the generated path and the obstacle detection information and, according to the behavior and the desired action for the robot, sends commands to the Control [18–20] module that moves the robot along the predefined path. Some of the low-level navigation procedures, which ensure that the robot is always in a safe state, include hardware fault identification, obstacle detection, and maximum velocity limitation.

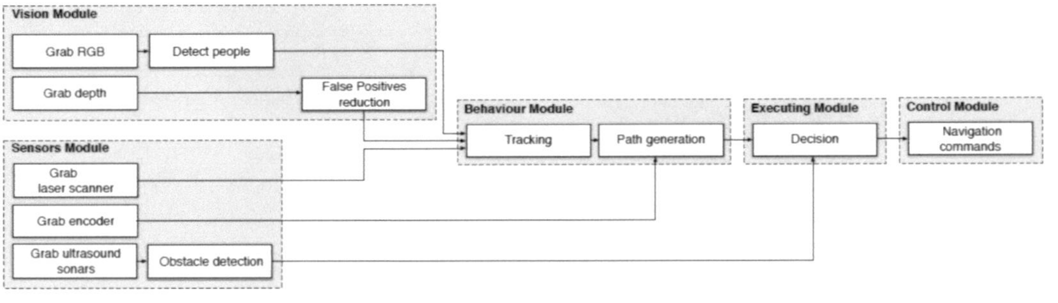

Figure 6. wGO software flowchart.

5. Results

In the first part of this section (subsection 5.1), some real scenario results are shown. A formal, quantitative, real-world evaluation is highly complicated due to many complex factors, such as the need to test in multiple different environments, testing with several user groups (including those with reduced mobility), the lack of any accepted standard evaluation protocol for the objective measurement of robotic assistance in a retail environment, etc. Therefore, only initial qualitative results based on realistic experimentation are shown in this paper. A formal evaluation which addresses each of these issues will be performed in future work.

The second part (subsection 5.2) describes a user satisfaction inquiry made on a real retail scenario on a population of 78 volunteers and its results.

5.1. wGO: Technical results

Starting with the detection process, the top left part of **Figure 7** shows the original RGB capture from a typical user following scenario in a commercial shopping environment. In the bottom left, the initial detection gives rise to two false positives, corresponding to the two ladies in the back, while the intended target is the men with his back to the robot. By using the RGB and depth information, shown in the top, these false positives can be removed, with the result shown in the bottom right.

The tracking process is illustrated next. In **Figure 8**, the person is visible both by the vision and by the 270° laser scanner, while in **Figure 9**, the person is only visible by the 270° laser scanner. In both cases, the tracking is not lost and the wGO can follow the person.

Path generation is used to decide about the navigation strategy of the wGO. Since the tracking module can in general return results from multiple sources of information (e.g., vision and laser), it is necessary to merge (fuse) them into one. An example of this result combination is provided in **Figure 10**. This fusion step makes the system more robust to errors in either one of the sensors and helps in producing more stable trajectories.

Figure 7. People detection example. Top left: RGB information; top right: depth information; bottom left: original detections; bottom right: result after removal of false positives.

Figure 8. Tracking example where the person is visible both by the vision and by the 270° laser scanner. Purple dot in the map image corresponds to the person as localized by the image module, while the blue dot corresponds to the person as localized by the 270° laser scanner.

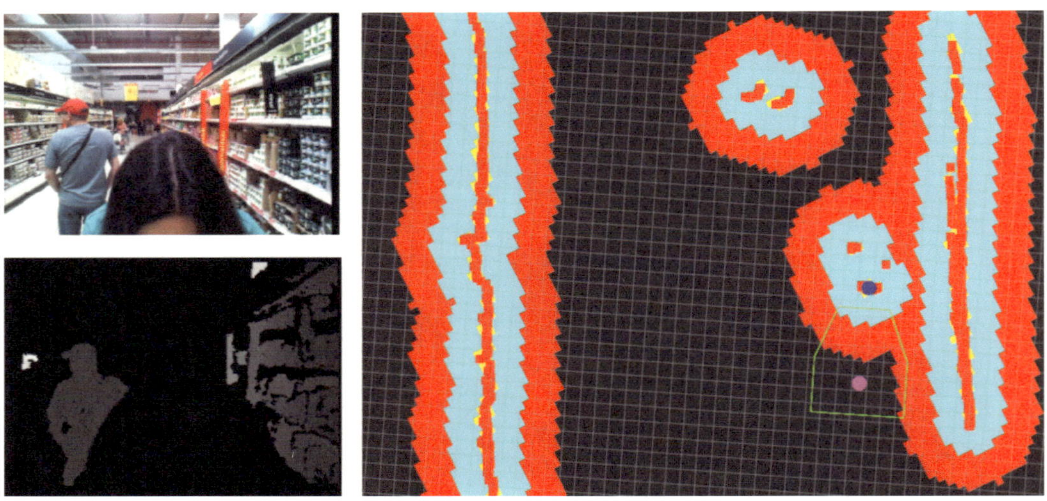

Figure 9. Tracking example where the person is only visible by the 270° laser scanner. Blue dot in the map image corresponds to the person as localized by the image module.

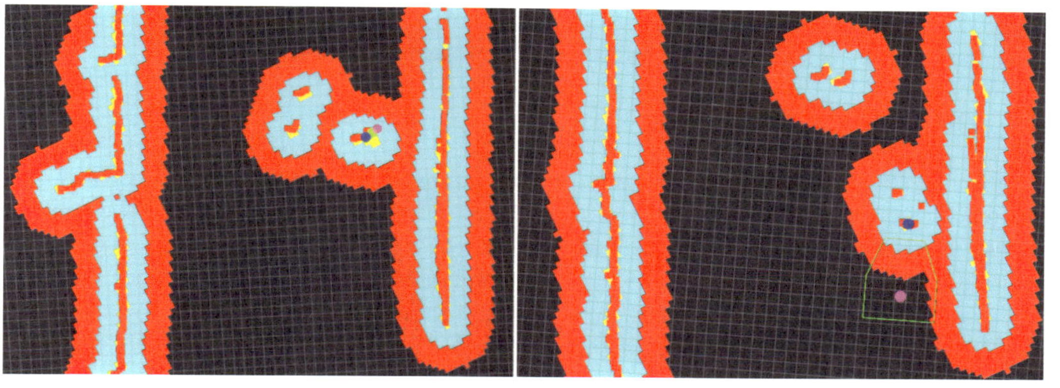

Figure 10. Fusion examples. On the left is a case where the person is visible both by the vision and by the 270° laser scanner, while on the right the person is only visible by the 270° laser scanner. Cross symbol in the map image of the left corresponds to the fusion result of the person as localized by the image module with the person as localized by the 270° laser scanner. Triangle symbol in the map image on the right corresponds to the person as localized by the image module and is also the final result.

Having one estimation of the person's localization, it is now necessary to decide where to send the robot (path generation). Moreover, it is important to keep some consistency in the results. Inaccuracies produced by the sensors can lead to highly unstable paths, which is not desirable. An example of a path made by the wGO is shown in the left part of **Figure 11**. An increase in the smoothness of the final route, when compared to a traditional approach, is observed.

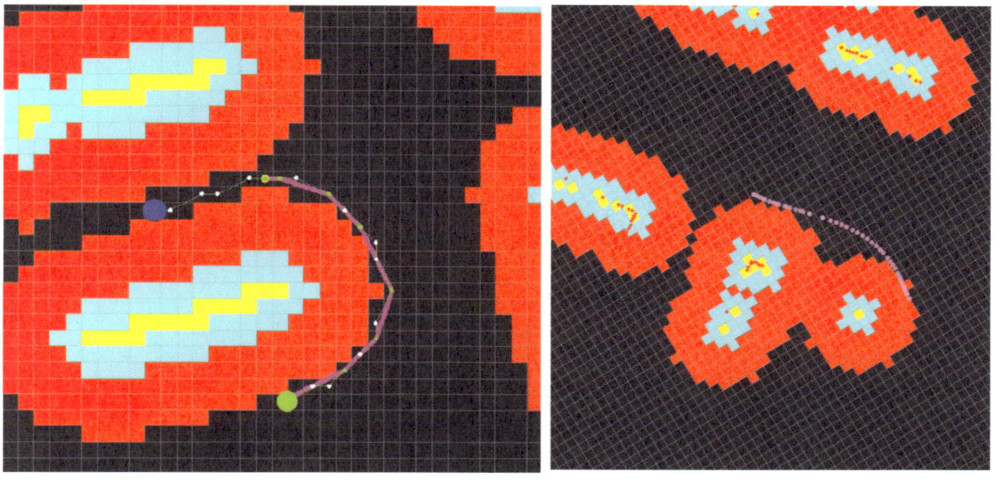

Figure 11. Path generation and effective path illustrations. In these figures, the purple dots are the effective path done by the wGO, blue circle represents the wGO, large green circle the target destination, white circles the waypoints generated by a traditional path generation approach, smaller green circles the waypoints given by the technique present in the wGO, red dashed lines the trajectory given by the traditional algorithm, and purple dashed lines the final trajectory produced by the wGO's system. In the background map, yellow and cyan areas are obstacles, red areas are security zones (although it is not advisable, the wGO can still use them if strictly necessary) and gray areas are free zones.

Finally, a sample of the control results is given in the right part of **Figure 11**. It can be observed that the trajectory is stable while avoiding all the present obstacles.

5.2. wGO: User satisfaction survey

To provide some qualitative feedback on the operation of the wGO, a user survey was conducted. The questionnaire was divided into six main blocks. The three first blocks contained statements designed as a Likert scale (with four levels in the first two blocks and 10 levels in the third one), the fourth block had an open-ended question, the fifth block had only one question concerning the shopping frequency, and the last block considered some demographic information. The survey is as follows:

1. Block 1: What is your satisfaction level with the following wGO aspects? (four levels Likert scale: "Highly unsatisfied", "Unsatisfied", "Satisfied" and "Very satisfied")
    - Available space to move in the store
    - Use in comparison with the other alternatives (e.g. scooters)
    - Speed of the shopping process
    - Commodity during the shopping process
2. Block 2: Future Use (four levels Likert scale ranging from "Unlikely" to "Very likely")
    - Would you use the wGO again?
3. Block 3: General satisfaction (ten levels Likert scale ranging from "Very bad" to "Very good")
    - How do you evaluate the wGO?
4. Block 4: Open-ended question
    - What would you change in the wGO?
5. Block 5: How often do you go to the supermarket?
    a. Once a month
    b. Twice a month
    c. Once a week
    d. Twice a week
6. Block 6: Demographics
    - Sex
    - Age

The survey was answered by 78 people, with significantly more females (49) than males (29). The age distributions are as illustrated in the left side of **Figure 12**. Note that there are

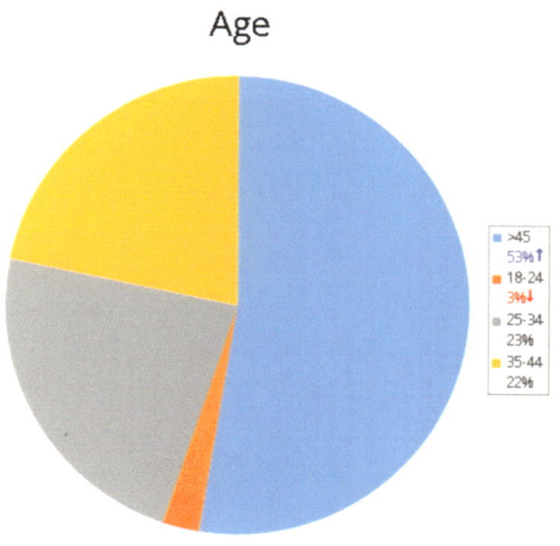

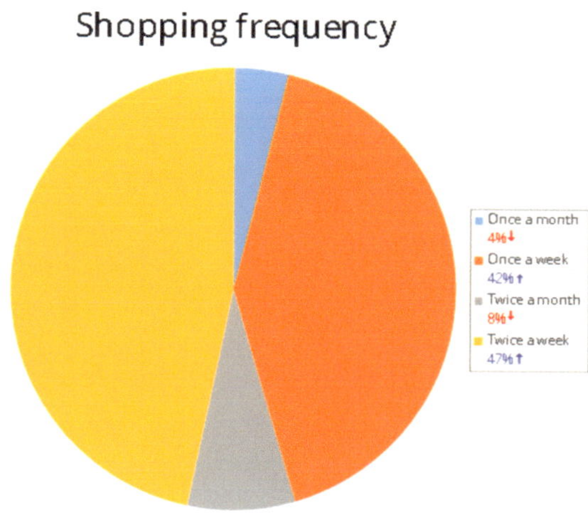

Figure 12. Age and shopping frequency distributions.

significantly more volunteers on the older age class (older than 45) and significantly less of the younger class (between 18 and 24). Also, in the age class of 25 to 34, most of the volunteers are females while in the class of people higher than 45 there is a significantly higher number of male users.

Considering shopping frequency, most of the survey participants go to the supermarket as often as once a month or twice week (right side of **Figure 12**). As a curiosity, all the "once a month" shoppers are male.

Results for question of Block 1 are shown in **Figure 13**. No volunteer attributed score 1 to any aspect, and people rated highly all the wGO aspects (with average scores always above 3.3). While its commodity was the highest rated aspect (being the amount of "Very satisfied" answers significantly high), its space in store was the lowest rated one (being the amount of "Very satisfied" answers significantly low and the amount of "Satisfied" significantly high), pointing to the fact that there are still some slight improvements to be made. People in the age class of 18 to 24 are significantly less happy with the space within store than other age groups. When looking at commodity, people in the same age group of 18 to 24 are mostly "Satisfied" (being this rate significantly higher).

When looking at the relationship of the shopping frequency with the speed, no "once a month shopper" was "Very satisfied" with the wGO speed. Conversely, "twice a week" shoppers were significantly "Very satisfied" with the speed, as illustrated in **Figure 14**.

Similarly, when looking at the relationship of the shopping frequency with the commodity, most of the "once a month" shoppers are "Unsatisfied," while "twice a week" shoppers were significantly "Very satisfied" (**Figure 15**).

Results for question of Blocks 2 and 3 are shown in **Figure 16**. People are very likely to use the wGO again (being the amount of "Very likely" answers significantly high). Within the "Very likely" future users, a significantly high number belongs to the higher than 45 age group.

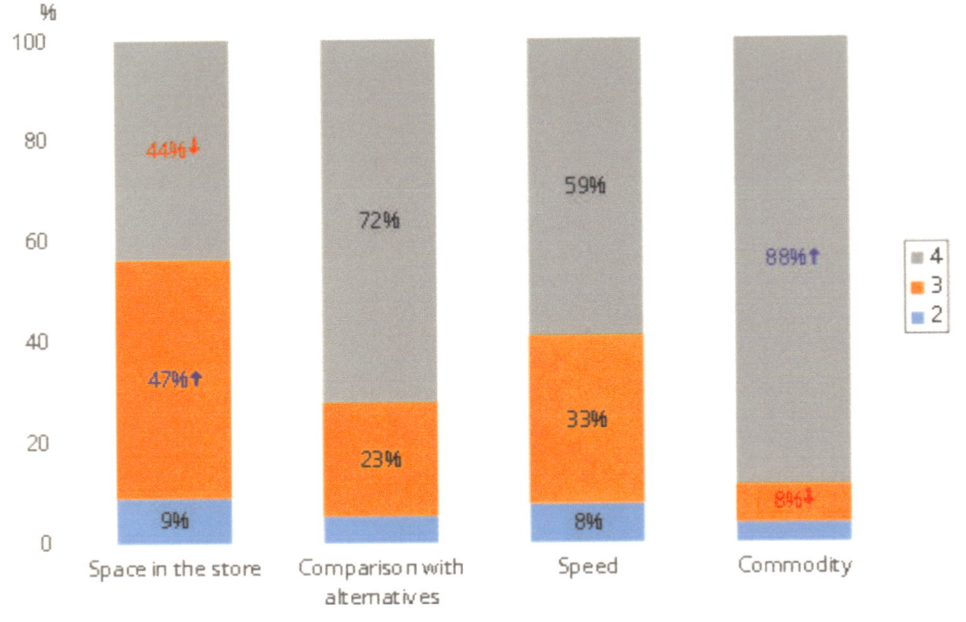

Figure 13. Results for Block 1: satisfaction level with some of the wGO aspects.

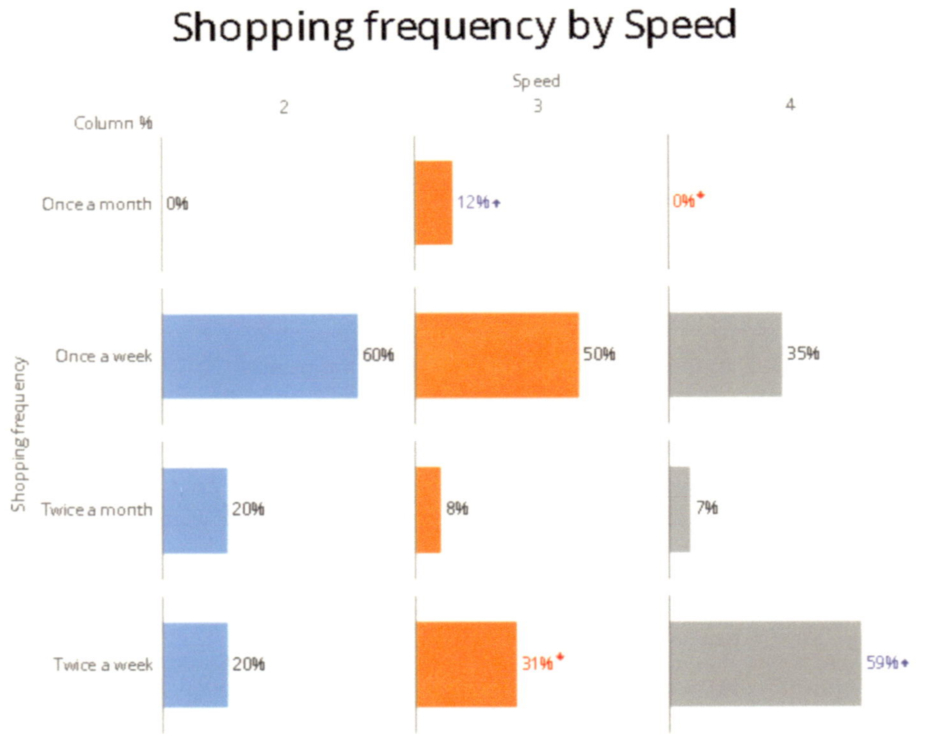

Figure 14. Shopping frequency versus wGO speed.

http://dx.doi.org/10.5772/intechopen.70277

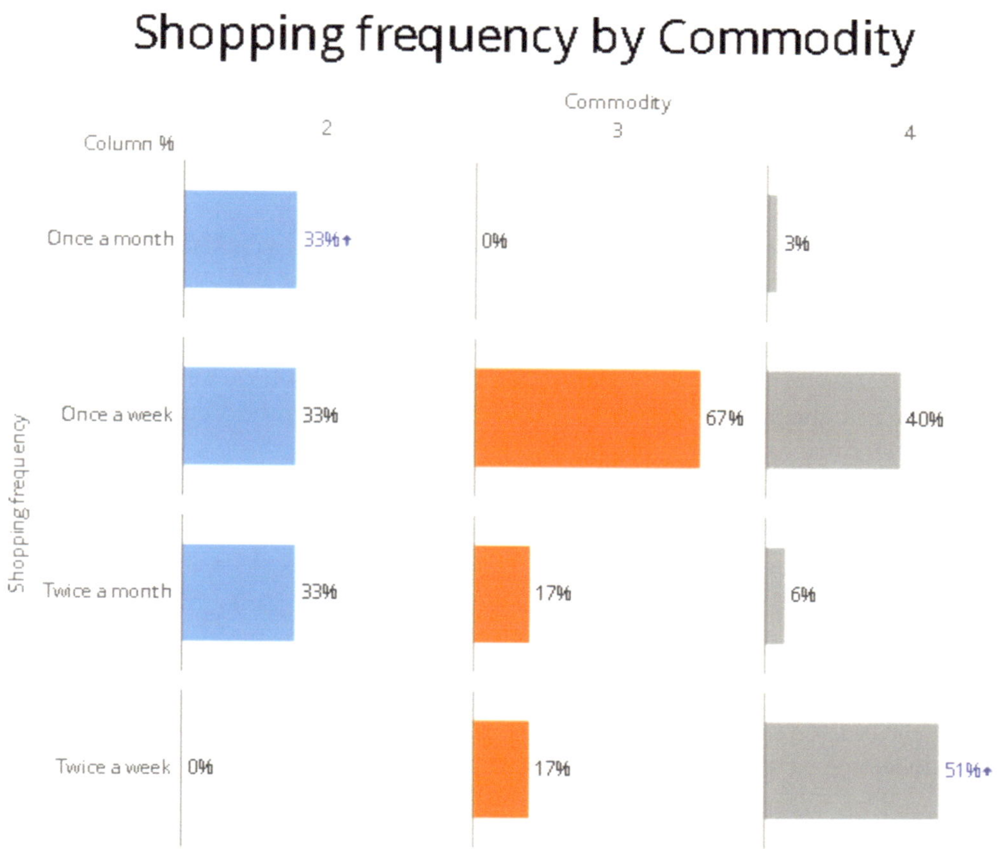

Figure 15. Shopping frequency versus wGO commodity.

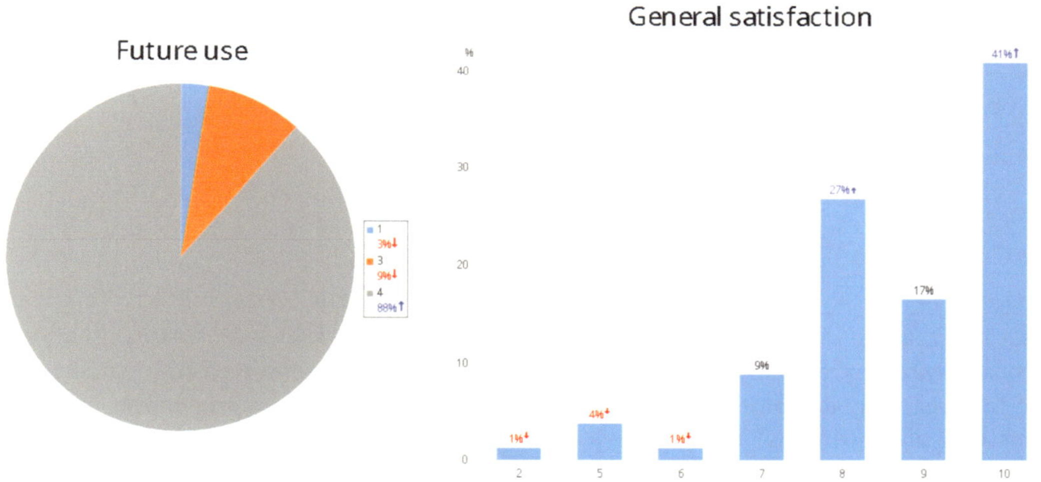

Figure 16. Results for Block 2: future use and Block 3: general satisfaction.

A significantly high number of people attribute the highest possible score on the general satisfaction scale.

When analyzing relationships between the several variables and the general satisfaction, it can be concluded that:

- User gender has no effect on general satisfaction.
- All the users in the age range of "18-24" gave a score of 8 to the general satisfaction question.
- "Once a month shoppers" attribute a value of 6 or 7 to the general satisfaction question.
- Among the "very satisfied" users with the Space in the store, 65% have "General satisfaction" of "10" (this is higher than the average for "10"). Conversely, among the "Unsatisfied" users with the Space in store factor, 29% have "General satisfaction" of "5" (this is higher than the average for "5").
- Users that believe wGO compare poorly with other alternatives and also give low scores in the general satisfaction question. The value of the Pearson Correlation Coefficient for this relationship is 0.6472, which translates into a moderate-positive correlation.
- Among the users that are "Very satisfied" with the speed, 24% have "General satisfaction" of "9" (this is higher than the average for "9").
- Less satisfied users with the wGO commodity also score lower on general satisfaction. The value of the Pearson Correlation Coefficient for this relationship is 0.6096, which translates into a moderate-positive correlation.
- Future use has a moderate-positive correlation with general satisfaction (Pearson Correlation Coefficient = 0.6014). Moreover, among the users who are "Unlikely" to use the wGO again, all of them have "General satisfaction" of "5" or less.

Concerning the open-ended question of Block 4, some limitations were pointed out, namely:

- limited space for the shopping items;
- low velocity;
- obstacle avoidance should be improved; and
- increase the robustness ("lose" the operator less times).

6. Conclusion

The wGO, an autonomous shopping cart, has been introduced. Experiments made in real scenarios are very encouraging, and a high-user satisfaction was observed. The participants on the user study demonstrated a comfortable behavior during the experiments as well as a very easy understanding of the robot's operating system (especially, related with the perception and navigation). Comments like "My shopping was very fast!", "In fact, it was a precious help!", and "I think it is awesome, I will certainly use and recommend it!" were made by the volunteers.

Some problems, however, remain to be solved. One of them is the limited space for the shopping items. Concerning the robot's behavior, low velocity and identification errors were mentioned.

Although the short-term application for the wGO is for commercial environment usage scenarios, several other applications are foreseen, for instance, at the shop floor of the manufacturing industry and logistics.

Acknowledgements

The authors would like to thank all the supporters, including TIC RISCO, Portugal Ventures, PME investiments, Finova, Compete, QREN, União Europeia Fundo Social Europeu, Mais Centro, Compete 2020, União Europeia Fundos Europeus Estruturais e de Investimento, sifIDe, inovaRIA, CEiiA, Fundão, UBI, Tente. We would also like to thank Portugal 2020, with the projects POCI-01-0247-FEDER-00.6398.

Author details

Joana Santos*, Daniel Campos, Fábio Duarte, Filipe Pereira, Inês Domingues, Joana Santos, João Leão, José Xavier, Luís de Matos, Manuel Camarneiro, Marcelo Penas, Maria Miranda, Ricardo Morais, Ricardo Silva and Tiago Esteves

*Address all correspondence to: joanasantos@followinspiration.pt

Follow Inspiration, Matosinhos, Portugal

References

[1] CRPG, Centro de Reabilitação Profssional de Gaia and ISCTE, Instituto Superior de Ciências do Trabalho e da Empresa. Elementos de Caracterizacção das Pessoas com Deficiências e Incapacidades em Portugal. In: Programa Operacional de Assistência Técnica ao QCA III: eixo FSE. 2007

[2] Government of Western Australia. Disability Services Commission - Disability Access and Inclusion Plan - Training Package - Section Five - Access and Inclusion [Internet]. Available from: http://ww2.health.wa.gov.au/Reports-and-publications/DisabilityAccess-and-Inclusion-Plan [Accessed: November 18, 2016]

[3] Rodriguez VJL, Fernandez LM, Hansen MM, Cortés EM, Peligero FJA, Torijano JL, Guallar EE, Cornudella BS, Verdejo IG. Shopping Cart. Justia Patent 20140265184. 2014

[4] Iezzoni LI, McCarthy EP, Davis RB, Siebens H. Mobility difficulties are not only a problem of old age. Journal of General Internal Medicine. 2001;**16**(4):235-243. DOI: 10.1046/j.1525-1497.2001.016004235.x

[5] Gurgel Pinheiro P, Cardozo E, Cláudio GP. Anticipative shared control for robotic wheelchairs used by people with disabilities. In: IEEE International Conference on Autonomous Robot Systems and Competitions. Vila Real, Portugal: IEEE; 2015. pp. 91-96. DOI: 10.1109/ICARSC.2015.26

[6] Faria BM, Reis LP, Lau N. User modeling and command language adapted for driving an intelligent wheelchair. In: IEEE International Conference on Autonomous Robot Systems and Competitions. Espinho, Portugal: IEEE; 2014. pp. 158-163. DOI: 10.1109/ICARSC.2014.6849779

[7] Figueredo M, Nascimento A, Monteiro RLS, Moret MA. Analysis of a sorter cascade applied to control a wheelchair. In: Gorrostieta-Hurtado E, editor. Robot Control. InTech; 2016

[8] Pavon-Pulido N, Lopez-Riquelme JA, Pinuaga-Cascales JJ, Ferruz-Melero J, Morais Dos Santos R. Cybi: A smart companion robot for elderly people: Improving teleoperation and telepresence skills by combining cloud computing technologies and fuzzy logic. In: IEEE International Conference on Autonomous Robot Systems and Competitions. Vila Real Portugal: IEEE; 2015. pp. 198-203. DOI: 10.1109/ICARSC.2015.40

[9] Martin Rico F, Rodriguez Lera FJ, Matellan OV. MYRABot+: A feasible robotic system for interaction challenges. In: IEEE International Conference on Autonomous Robot Systems and Competitions. 2014. pp. 273-278. DOI: 10.1109/ICARSC.2014.6849798

[10] Faria V, Silva J, Martins M, Santos C. Dynamical system approach for obstacle avoidance in a smart walker device. In: IEEE International Conference on Autonomous Robot Systems and Competitions. Espinho, Portugal: IEEE; 2014. pp. 261-266. DOI: 10.1109/ICARSC.2014.6849796

[11] Gómez-Goiri A, Castillejo E, Orduña P, Laiseca X, López-de-Ipiña D, Fínez S. Easing the mobility of disabled people in supermarkets using a distributed solution. In: Bravo J, Hervás R, Villarreal V, editors. Ambient Assisted Living. Lecture Notes in Computer Science. Berlin, Heidelberg: Springer; 2011. pp. 41-48

[12] Jafari OH, Mitel D, Leibe B. Real-time RGB-D based people detection and tracking for mobile robots and head-worn cameras. In: IEEE International Conference on Robotics and Automation. Hong Kong China: IEEE; 2014. pp. 5636-5643. DOI: 10.1109/ICRA.2014.6907688

[13] Nebehay G, Pflugfelder R. Clustering of static-adaptive correspondences for deformable object tracking. In: IEEE Conference on Computer Vision and Patern Recognition; 2015. pp. 2784-2791. DOI: 10.1109/CVPR.2015.7298895

[14] Haro F, Torres M. A comparison of path planning algorithms for omni-directional robots in dynamic environments. In: IEEE 3rd Latin American Robotics Symposium. Santiago Chile: IEEE; 2006. pp. 18-25. DOI: 10.1109/LARS.2006.334319

[15] Abdulhafz WA, Khamis A. Bayesian approach to multisensor data fusion with pre and post-fltering. In: 10th IEEE International Conference on Networking, Sensing and Control. Evry, France: IEEE; 2013. pp. 373-378. DOI: 10.1109/ICNSC.2013.6548766

[16] Government of Western Australia. Disability Services Commission - Disability Access and Inclusion Plan - Training Package - Section Five - Access and Inclusion [Internet]. Available from: http://ww2.health.wa.gov.au/Reports-and-publications/DisabilityAccess-and-Inclusion-Plan [Accessed: November 18, 2016]

[17] Einicke GA, editor. Smoothing, Filtering and Prediction—Estimating the Past, Present and Future. Croatia: InTech; 2012. DOI: 10.5772/2706

[18] Araki M. PID control. In: Control Systems, Robotics, and Automation- Vol. II. Encyclopedia of Life Support Systems (EOLSS). Japan: Kyoto University; 1996

[19] Palmieri L, Arras KO. A novel RRT extend function for efficient and smooth mobile robot motion planning. In: IEEE International Conference on Intelligent Robots and Systems; 2014. pp. 205-211

[20] Park JJ, Kuipers B. A smooth control law for graceful motion of differential wheeled mobile robots in 2D environment. In: IEEE International Conference on Robotics and Automation; 2011. pp. 4896-4902. DOI: 10.1109/ICRA.2011.5980167

Chapter 2

User Experience Results of Setting Free a Service Robot for Older Adults at Home

Markus Vincze, Markus Bajones, Markus Suchi,
Daniel Wolf, Lara Lammer, Astrid Weiss and
David Fischinger

Additional information is available at the end of the chapter

http://dx.doi.org/10.5772/intechopen.70453

Abstract

The chapter presents the analysis of user trials where, for the first time, a service robot was set free in the home of users. Different to previous studies there was no pre-specified schedule of tasks to execute. The goal was to show that useful functionalities for users can also be achieved with the low-cost components of the Hobbit robot. With the one-arm mobile service robot Hobbit we provided users with a service robot running basic robot functionalities such as navigation, grasping objects from the floor, emergency handling, entertainment, fitness and communication functions. Users could freely select what to do over the three-week trials in homes in three European countries. Users have been questioned on what functionality would help them to stay longer at home and live independently. Results provide better insights of what users want than in pre-set scenarios, where many of the factors we encountered do not show up. Good examples are the need to have robots navigate autonomously at home, grasping objects from the floor is a highly valued function, and the robot needs to adapt locations depending on the daily liking of the users who move much more freely at home than in pre-set scenarios.

Keywords: service robots, personal robots, autonomous navigation, grasping, user experience at home, object recognition

1. Introduction

Robots have long been a dream of humans as helpers. We started from the vision of a robot helper at home. While there may be a large literature and demand for robot to help all people in particular with household chores, too many of the chores are not yet possible to technically realise. Hence, we rather considered present robot technical capabilities and started to envision the role of a home robot enabling older people to feel safe and stay longer in their homes. The result of two iterations of user-centred design is the Hobbit robot. **Figure 1** shows the robot and lists its main components, which will be introduced in more detail later.

The rationale behind this selection of feeling safe at home is that when an older adult falls it is necessary to transfer to a care facility. Since this is the primary cause for these transitions, any measure to increase the perceived safety at home will aid to improve the situation for the users. Consequently, we introduce a mobile robotic solution that sets out to discover falls. It does so in a proactive way to avoid falls in the first place. The concept has been developed together with professional care personnel. The result is the robot Hobbit. It provides a set of

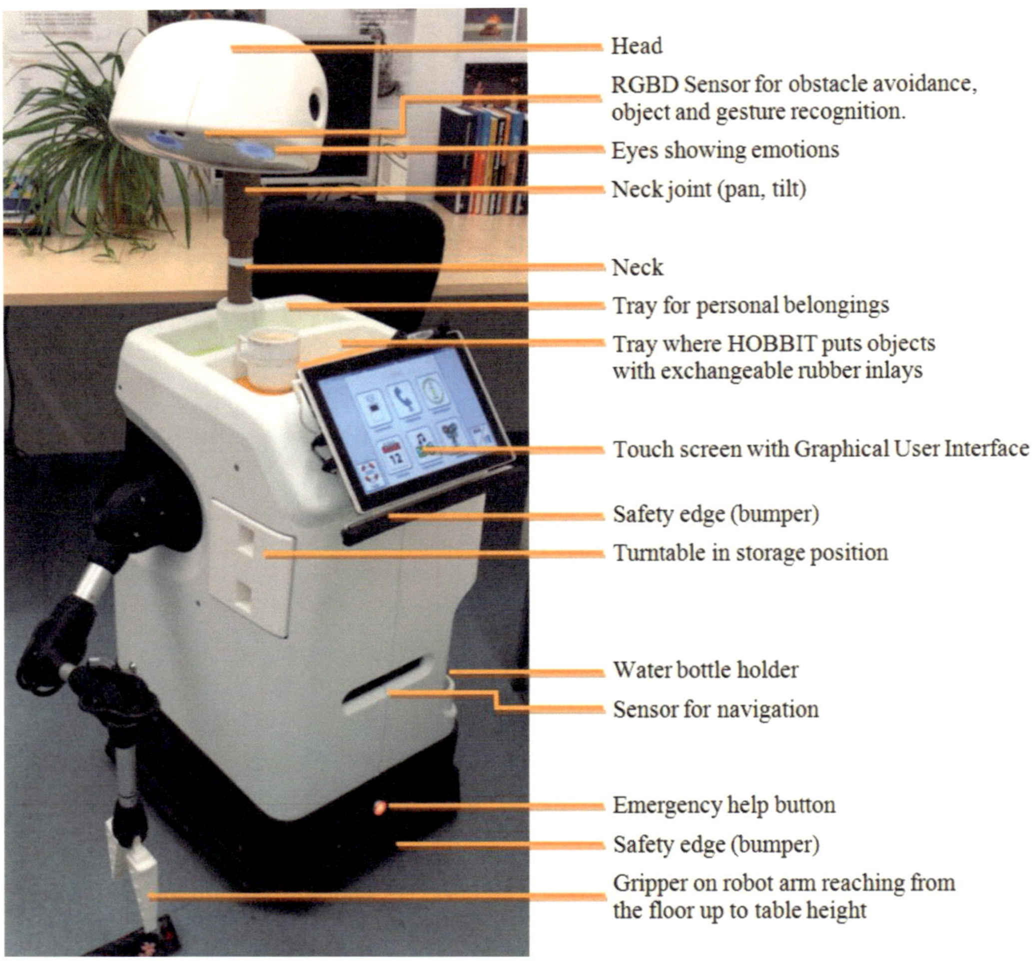

Figure 1. The service robot Hobbit designed to help older adults stay longer independent at home. Its primary functionality is coping with emergency situations, grasping objects from the floor or transporting objects to avoid falls and is a good collection of entertainment and physical and cognitive fitness functions.

functionalities to the older adults. It uses multimodal interface based on speech, gestures and touch-screen for realising easy-to-use human-robot interaction (HRI).

Today, there are several service robot solutions targeting the application of an extended video phone. However, these do not include an arm to actively interact with objects in the user home. The highest developed robot with similar capability as Hobbit is Care-O-bot. It has been tested in many trials in care facilities for studying user interaction, for example, for bringing water or other assistive operations in care facilities [1]. However, this robot is too large to operate at homes; it would hardly fit through doors. Many of the remaining mobile service robots, for example, Giraff, are not able to autonomously navigate and need to be operated remotely.

The main novelty was to bring a service robot into user homes that can do more than a video phone on wheels that is operated by a remote user including the navigational capacity. The designed Hobbit robot provided to the user the following functions:

- Maintaining the user's self-efficacy is addressed with exercising cognitive and physical skills (social connectedness and fitness functions).
- Increasing the perceived user safety is addressed by managing a safe home including functions such as emergency detection, grasping from the floor, transporting objects, patrolling to check for the user and calling the robot.
- Positive affect towards the robot is addressed using entertainment functions. This also increases the user's well-being.

A unique setting of the Hobbit study was that users were free to select any of the functions at any time.

In this contribution we report the findings of the user experiences with the Hobbit robot. The trials involved 18 users in 3 different countries spanning from the north to the south of Europe: Sweden, Austria and Greece. Since the users wdid not have any given schedules or scripts, they were free to use the robot as they wanted. The idea was to set the robot free to better find out what the users would really want from the robot rather than presenting the users with a fixed script or setting as in previous studies.

The chapter proceeds as follows. After a review of related service robots for elder care, we present results of a study on what the robot should be able to do in a safe home scenario (Section 3). Section 4 presents the robot and its components and Section 5 presents the results of the user study.

2. Related work

Service robots for older adults are typically aligned according to their capability to address activities of daily living (ADL) or instrumental ADL (IADL). However, typical functions of ADL/IADL are dressing, food preparation, eating, cleaning and rehabilitation or direct physical exercise activities. All these functions are very difficult to realise, and there are only a few robots in research settings progressing towards one of these functions. Hence, we took another approach to realise a useful robot. Coming from the fundamental need of feeling safe at home, we introduced a series of functions to maintain safety at home and augment this with functions to entertain and motivate the user. The intention is to create a positive effect with these socialising functions.

Studies can be compared on where user trials have been conducted. Indeed, very few studies go beyond tests in professional care facilities, and even fewer study longer durations, for example [2] and a recent survey in [3]. In the following we highlight recent developments. Today, many service robot projects further advance one specific functionality. For example, in GrowMeUp (http://www.growmeup.eu/), the user's habits, preferences and routines are studied using multiple sensors on the robot and the environment. The robot is a reduced PAL platform without arms from Pal Robotics (http://pal-robotics.com). As we have seen in our user experiences, relying on an external sensor network may be feasible in specially designed homes but is not welcome by users at home and requires substantial installations and the related costs.

The EU project EnrichMe (http://www.enrichme.eu) also studies the use of touch-screen and augmented user interface as a follow-up of the CompanionAble Project with Robosoft and the Kampai robot. Ambient-assisted living (AAL) functions are used by introducing radio-frequency identification (RFID) chips into objects. Project Mario (http://www.mario-project.eu/) addresses isolation, loneliness and dementia of older adults through multi-faceted interventions delivered by service robots including the use of AAL installations. The project partners use a near state of the art platform based on the Pal Robotics robot that is 'flexible, modular friendly, low cost and close to market ready in order to realise field contributions in the immediate future'.

An aspect that has to be kept in mind is that the generation of adult 70+ users included in the developments here differs from future older users in regard to experience with and acceptance of technology. Currently, literature discusses the so-called digital divide between people over 70 and younger users. This digital divide is shrinking (cf. [4]). This means it can be assumed that future older users will face less difficulty in using, for instance, web browsers and other computer tools than the older generation of today. In fact, Hanson argues that older adults form the fastest-growing group of web users. In other words, future users may be more critical with the interface but less afraid of using new technology.

On the other side, age-related functional limitations affecting interaction with computers and other interactive devices will very likely stay the same for future generations of older users. These include cognitive changes with regard to short-term memory, concentration and also solving of new types of problems, as well as common perceptual impairments (sight and hearing), and finally reduced motor skills. Such perceptual and motor impairments are taken into account for the design of a user-friendly multimodal user interface (MMUI) that is to be evaluated in the user trials (based on experience from other projects with older users, such as KSERA).

The contribution of the work presented here is a robot that acts fully autonomously in the home of the user - the Hobbit robot. As pointed out in this review, up to today robots have rather been remotely operated, or a small number of tasks in the user home have been studied in the user tests. Examples are projects such as Giraff++, SRS and Robot-Era. Some of the coordinators of these projects pointed out that these robots urgently need the capability of autonomously navigating in the user homes. And exactly, this is the capability developed within the Hobbit project, and it will be of good future use when moving towards longer-term tests at home.

3. Requirements of a home robot for increasing the perceived safety of older adults

Studies in service robotics repeatedly reported the requirements humans have to robots. The top are the well-known four Cs for cleaning toilet, bath, kitchen and windows. This need is confirmed even in our user trials. An older lady saw the robot and immediately responded with 'I do not need this robot, it cannot clean windows'. This clearly indicates that this is a rather large gap between what users in general, not only older adults, would want from a robot and what robots are actually able to do today.

Conscious of this gap and of what robots can actually perform today and what users would want, one of the motivations for running the Hobbit robot in the user homes was to get a better understanding of what services the robot could actually provide.

Consequently, we started from functions and services that a robot *could* provide to users. In a first study to investigate what older adults (primary users (PUs)) would need at home, we conducted a questionnaire with questions regarding the functionality, safety and operation of a home robot. The user consisted of 113 persons with an average age of 76.2 years. Overall, 69 (61.1%) of this group were female and 43 (38.1%) male. Forty-six (40.7%) of the primary users were single-living, whereas the remainder stated to live with 1 or more persons in the household; 18 people did not answer the question. The majority of interviewees lived in a flat (59.3%), 24.8% in a house, 14.2% in a nursing home. Two persons did not answer the question. Most PUs (62.8%) did not receive any home help service, healthcare service or support from relatives. Only six (5.3%) interviewees were permanently living with relatives or a care service. Asked about the frequency of using a computer, 49 persons (43.4%) stated not to use computers at all and 38 (33.6%) to use computers every day. This balanced amount of 'computer literacy' makes for an even sample, since many potential purchasers of Hobbits on the market cannot be expected to have experience with handling computers.

Tables 1–3 summarise the most important results of the frequency analyses for the questions within the group of primary users (n = 113). Sample size varies for each item, due to varying numbers of answers. Not every participant answered every question. **Tables 1–3** depict the above average percentages of 'agree' answers to the questions. Most users wanted their robot to be able to search and find things, grasp objects from the floor and from a shelf and also bring objects to them (**Table 1**). Other important functions were reminding users of appointments or phone calls and their medication.

Question	Valid sample	Percentage
Search and find	107	86.9
Grasp from the floor	109	86.2
Grasp from the shelf	105	80.0
Fetch and bring	106	79.2
Reminder (appointments or phone calls)	106	78.3
Reminder (medication)	107	77.6
Carry objects	104	68.3
Follow	104	66.3

Table 1. Most wanted functionalities of the service robot.

The majority agreed that they felt safe, if the robot could call for help (see **Table 2**). About 61.8% of the valid sample agreed that they wanted their robot to be active at night; this percentage becomes even stronger when those who chose 'I rather agree' are also added. The accumulated frequency then reaches 81.8%. Only 8 of 112 PUs (7.1%) stated that they felt frightened by the idea of having a robot at home.

The idea of having the robot taking care of its own battery level is highly popular (96.4% of 112 answers). On the other hand, users like to stay in control. This is reflected in the high amount of 'agree' answers to the statement that the robot can only do what it is told by the user. This topic has to be considered carefully when designing autonomously triggered activities of the Hobbit system (e.g. reminders). Operation of the robot should be preferably speech based. Remote control is in the second place, followed by gestures and touch-screen (see **Table 3**).

3.1. The functions provided by the Hobbit robot

The results presented herein mark a significant step forwards in evaluating robotic systems under real-life conditions [5]. For reasons of completeness, we shortly set the above presented functions in relation to state-of-the-art robots (see also [6]).

The selections of functions that have been implemented on the robot have been extracted from multiple interactions with users, secondary users or relatives and professional caregivers. We conducted first home trials with an autonomous robot with the aim to find out what users want.

Question	Valid sample	Percentage
Safe because of call for help	112	85.7
Active at night	110	61.8
Frightening	112	7.10

Table 2. Most wanted safety aspects of the service robot.

Question	Valid sample	Percentage
Self-charging system	112	96.4
Interaction through speech	102	88.2
Operation by speech	96	85.4
Do what I ask for	113	72.6
Operation by remote control	97	68.0
Move everywhere	108	65.7

Table 3. Most wanted modes of operating the service robot.

Here, a lot more work is needed, and recently started projects will expand our understanding. Part of this work was that we conducted two iterations of user studies and collected user requirements [7]. These requirements give a clear picture of what older adults would want at present from a robot helper at home. Conclusions are drawn from workshops with older adults that created a longer list of requirements that have then been ranked in studies and questionnaires and correlated with technical feasibility given the present state of the art in service robotics. We used first user trials and lessons learned to verify the ranked requirements [8].

Before reviewing the robot system concept (Section 4), we summarise the user requirements and relate them to other studies or care robots. The clear requirements formulated within the Hobbit project still hold. The main services that a home robot should provide to aid older adults target the following needs. Note that the items listed below are the convergence of functionalities that can actually be provided by the robot and the results of the questionnaires given in **Tables (1–3)**:

- Maintain the user's efficacy level: this includes functions for keeping an active and fulfilled live and includes:
  - Social connectedness includes telephone and Internet access to alternative ways of communication such as a video call or to access weather, news and other information.
  - Physical and cognitive fitness includes physical exercises that have been considered on top of the initial description of work. This includes games and playing music or video or radio, a function that has been surprisingly welcome by users to play the favourite radio station.
- Increase the perceived safety of the user:
  - A main aspect is already the physical presence of the robot and its care functions such as seeking the user and user interaction during the patrolling function.
  - Multimodal interaction capabilities and several ways to trigger an emergency call.
  - Pick-up of known and unknown objects from the floor which turned out to be an essential aspect. The normal scepticism towards the robot went away after seeing the robot picking up an object from the floor. A clean-up function further extends this capability.

- The robot provided an additional safety check with the user, making her/him aware of hazards at home while proposing solutions or options to assist.
- Calling the robot for help: the use of call buttons is an effective means to call the robot for any task at any time.

- Functions for the user's well-being: here, we summarise services that are nice to have and will actually assist to accept the platform and keep it in use. In the Hobbit idea, we had drawn out many of these functions as elements to make the user feel good and possibly even create a bonding to the robot such that it is trusted and used and the previous two aspects are reached to an even better degree. Examples of these functions are:
  - A first personalisation of the robot is executed in the initialisation phase. The users set initial parameters, which could be changed later. Additionally, the robot and basic guidelines for operating it are introduced.
  - Learning new objects and findings these objects are welcome features for the users and regarded as a great commodity.
  - An important functionality that extends the functionalities provided in Hobbit is the pick-up from high locations. Grasping objects from places high up that cannot easily be reached will be investigated in EU project RAMCIP, though robot costs are expected to be considerable higher. In Hobbit we regard this functionality as a future module and a possible extension of the basic robot platform.
  - Entertainment ranges from games over music to surprising the user. All these functions aim at increasing the user acceptance.
  - Reward functionality is a means to enhance the user binding with the hypothesis to improve the acceptance by the user.

In summary, the Hobbit robot provides a rich repertoire of functions, where several are novel and have been tested with users or at home for the first time.

4. The components of the Hobbit robot

The main components of the robot have already been depicted in **Figure 1**. The robot platform used for the home trials has differential drive kinematics developed by MetraLabs: a floor-parallel depth camera for purposes of navigation, a head mounted 120 cm above the ground to the level of the height of a sitting person, a touch-screen mounted at an angle in front of the torso or the robot and on the right side of the robot a manipulator to pick up objects. The head contains two screens to present the robot's eyes and an RGBD camera (ASUS Xtion). This configuration is an improvement of the previous Hobbit version and implements the lessons learned in a series of previous user trials (see also for details [9]). The main dimensions have been reduced to follow user requests. The height of the Hobbit robot is now 125 cm, and it has a width of maximum 56 cm at the point where the shoulder of the arm sticks slightly out beyond the robot. Other features will be discussed in Section 5 when discussing the hull or individual features of the upper body and the head.

A key element of the development of the Hobbit robot set out to reduce the costs of the hardware costs to a minimum. For example, laser sensors are rather expensive and only operate in one plane. Replacing them with RGBD camera has the advantage that their cost will be lower and they provide full 3D perception. Hence, we can test the feasibility to cope with all the functionality needed at home and with lower price to reach closer to the expected costs of presenting a robot for home robotics. The hardware components sum up to 16,000 Euro. **Figure 1** presented the Hobbit robot with its main components. Navigation is autonomous and uses virtual laser scans from RGBD images. The robot operates using a multimodal user interface (MMUI) that comprises a graphical user interface (GUI) with touch-screen, automatic speech recognition (ASR), text to speech (TTS) and gesture recognition interface (GRI). The robot has functions for edutainment (music, radio, audiobooks, pre-installed web radio and services, games and cognitive fitness functions), reminders, video phone service, control of a manipulator, access to an ambient-assisted living (AAL) environment (e.g. call buttons) and emergency call features. The robot's functionalities included automatic emergency detection (e.g. patrolling and detecting persons lying on the floor), handling emergencies (communication with relatives) and supportive fall prevention measures (transporting small items, picking up objects from the floor, searching for objects the robot had been taught by the user).

5. Results of setting free a service robot for older adults at home

In the following we structure the results into the aspects regarding the robot usage (usability, acceptance and affordability) and issues related to the robot hardware, software and development. Before presenting the results, we summarise the design and methods used to evaluate the user trials.

5.1. Design and methods of the user trials

The user trials have been conducted in three countries, and users tested the robot for 3 weeks each. The trials took place in Austria with seven end-users, in Greece with four end-users and in Sweden again with seven end-users. In total, the trials included 18 primary users (PUs) and 16 secondary users (SU). The trials were carried out in the user homes with the robot interacting autonomously for 3 weeks with the user. All trials took place in private homes of single-living senior adults. Each trial with one user lasted 3 weeks. In total, the robot was deployed for 372 days. Assessment by means of qualitative interviews and questionnaires took place at four stages of each trial: pre-trial, midterm, end of trial and posttrial (i.e. 1 week after the trial had ended). Results of the qualitative interviews as well as perceived safety measured by the falls efficacy scale (FES) [10] are reported. Eighteen elderly users participated in this study, and 16 (14 female) were included for statistical analysis (two participants had to be excluded because of missing data). The mean age was 80 years, ranging from 75 to 89 years. Qualitative data were organised using NVivo (QSR International). Quantitative data were analysed using SPSS by means of descriptive statistics and non-parametric methods (Friedman ranking test).

We used a multi-method approach for testing the most important evaluation criteria: (1) usability; (2) acceptance, which includes the mutual care (MuC) concept [11]) and (3) affordability.

This testing followed an intricate evaluation procedure with regular updates using the inputs of the reviewers of the project and the first experiences of the pilot user tests [9]. The method mix used contained interviews; questionnaires; cultural probing with the older adults before, during and after the trials; and the continuous logging of all interaction data in the Hobbit robot (see **Figure 2**). Qualitative, quantitative, cultural probing and logging data were preprocessed and analysed according to state-of-the-art scientific rules and procedures. Detailed results of the field trials will be reported.

5.2. Results regarding overall robot usage

As outlined above, results were gained from questionnaires; interviews; cultural probing with the participants before, during and after the trials; and continuous logging of all interaction data in the Hobbit robot. Qualitative, quantitative, cultural probing and logging data were preprocessed and analysed according to state-of-the-art scientific rules and procedures. Detailed results of the field trials will be reported. The most important results of the user trials related to the three main quality criteria were:

- Usability: users agreed that Hobbit is easy to use and intuitive to handle. The option to use different input modalities was perceived as very helpful for PUs. There was, however,

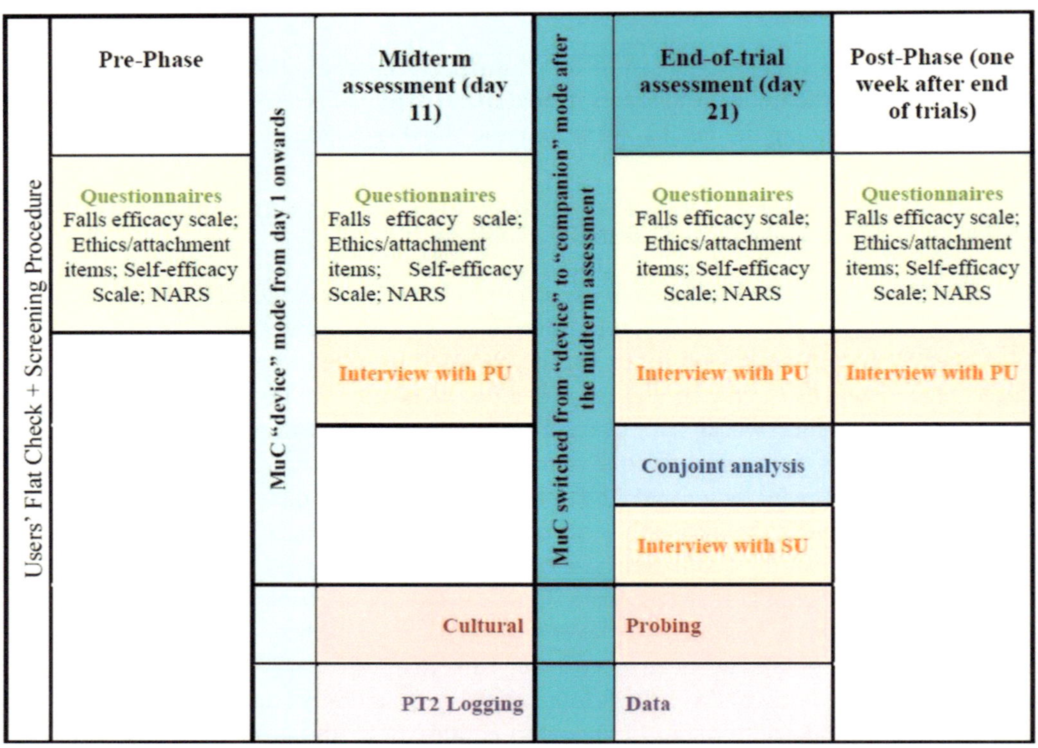

Figure 2. Overview of trial procedure and evaluation materials.

some lack of functionality, since not all functions worked all the time. This was acceptable for a prototype but obviously needs to be improved.

- Acceptance was ambivalent among users. In general the attitude towards the robot was positive and did not change. The emotional attachment weakened over the duration of the trials, mainly due to the technical problems. This also indicates that some of the expectations of the users could not be fulfilled. A more important finding is that the reciprocity was not perceived by the primary users. This indicates that the mutual care approach needs some refinement to become effective.
- Affordability: The results of the user trials indicate that the Hobbit robot is with its current price of 16,000 Euro—not yet affordable for the target users. While this is more than a magnitude cheaper than other similar robots, it is too expensive. On the other hand, a robot arm would be cheaper but is not wanted. Only a complete Hobbit robot with a pan/tilt head, a manipulators and functions for pick-up and for learning objects will all be valuable for PUs.

Using the qualitative data of the user trails, we can obtain insights on how the users like the different functions Hobbit provides. The users mostly appreciated the function to pick up objects, where picking up from the floor has rates with the highest value. Other highly welcome functions are detecting potential emergencies (adding to the feeling of perceived safety), the transport of objects, the cognitive and physical fitness functions and to be present to the user reminders. Although the pick-up function sometimes had failures and only worked without any difficulties for 18% in total of 372 days, users still saw the high potential. If this function would be available, users would want it. All in all, speed of operation of the robot system is not as good as it should be. Neither voice commands nor gestures operated to the expectations of the users. Consequently, the touch-screen has been used more often. To conclude, quantitative data shows that the perceived safety for the users, which is obtained using the Falls efficacy scale (FES) measure, did not increase along the user tests ($p = 0.265$).

Finally, the mutual care (MuC) concept, which has been proposed to foster the acceptance and improve the use of robot, has been detected to have rather little consequence. A cause for this unexpected result may be with high probability that the technical functioning of the robot is not yet high enough. Hence, there is considerable work ahead. While the trials indicate a first proof of concept, there needs to be more added reliability of all the robot operations. The good finding is that navigation is autonomous and with a service rate of over 98% was rated as sufficient by users. Hence, we provided a very good start for gaining acceptance by the users. Let us now have a closer look at the next step of technical improvements.

5.3. Results regarding technical improvements towards future home robots

A general conclusion is that robots such as Care-O-bot or Giraff are too big to be operated at homes. Users expect to be sitting when interacting with the robot. Consequently, robot height should be about a sitting height of a person. Ideal is about 120 cm. So, Hobbit is still 5 cm too high but comes close to the ideal. Interestingly, the only robot so far at this height is Pepper of SoftBank Robotics.

Home robots may first be used in the homes of older adults. In these envinronments many things have been accumulated over the years, hence space is sparse. Ideally, the robot footprint does not exceed a diameter of 40 cm, which is technically difficult to reach including an arm. The shape of the robot may be rectangular; however, to simplify the robot's navigation capabilities, our findings propose a cylindrical shape. This will make rotating behaviours simpler and allows an easier navigation through doors, which have been found to be as narrow as 60 cm, in particular in Sweden. Finally, a recommendation is to build the robot manipulator within the hull of the robot. It then will not collide with furniture or doors and walls when the robot is rotating or moving in the apartment.

A difficulty in many homes is not even floors and thresholds. Thresholds as high as 25 mm were encountered many times. We resolved the issue temporarily with ramps. However, it would be better to devise an efficient solution based on wheels to surpass small thresholds. Additions to the apartment are not always welcome by the users. Some examples of coping with thresholds are shown in **Figure 3**.

The battery duration of the Hobbit turned out to be sufficient. It lasted for about 8 hours of continuous robot uptime. A user will not use the robot this long. Users got tired after about 2 hours and sent the robot to go to the recharge station. Furthermore, for reasons of safety, the robot must always have a battery level of at least 30% to make sure that it can patrol the user home, locate the user and if necessary has enough power to initiate and operate through a full emergency scenario. The present implementation based on voltage was not ideal to reliably report remaining battery status. The battery status was not transparent for the user. Ideally, it should estimate the remaining run time.

Another more practical issue is the docking station. It needs to be small, since there is not much space in the homes of older adults, as stated above. It may contain markers for improving robust docking behaviour, but we simply used a trapezoid shape, which turned out to be sufficient. **Figure 4** depicts the docking station.

Another practical issue is a way to stop the robot at any time. While we use a hardware on/off button, a practice showed that there should also be a 'cancel' hardware button or switch that would allow the user to cancel immediately whatever activity the robot is doing at any given moment. The advantage of this button is also that this could be done remotely so that the user does not have to approach the robot. This will be a simple add-on and put the user in charge of the robot at any time.

Figure 3. Several examples of mats and ramps to make the floor flat enough for a wheeled robot.

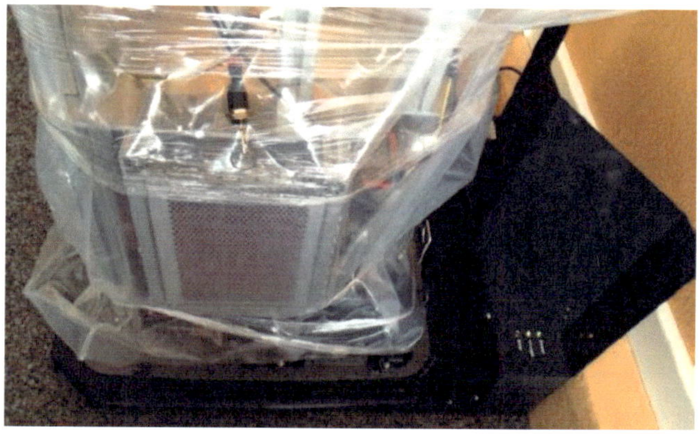

Figure 4. Docking station (black) and the robot wrapped in plastic for temperature tests.

5.4. Robot safety edges and bumpers

Hobbit was equipped with safety edges (see **Figure 5**) at the base plate and the tablet's base plate. They are electrically interconnected with the main electronics forcing the Hobbit robot to stop when it is colliding with obstacles or humans. In this case the electro-conductive inner sides of the safety edges are pressed together, and the resulting signal leads the main electronic to stop the drive motors.

The main bumper (**Figure 5**(left)) is at the floor and also assures that the user cannot get under the robot, e.g. with the foot. However, in a few cases, shelves would protrude just above the bumper. To address this it would be helpful to design a second bumper at 20 cm height so kitchen cupboards with cabinet plinths are not in a risk to be hit. For a commercial product, a low-cost solution for the bumper needs to be determined.

At present the robot has no sensors pointing backwards; only the bumper goes all way around. However, an infrared or sonar sensor such as additional sensor and further forward-looking sensor could be added. With the additional sensor at a height of 20 cm and a better localisation and mapping method, this issue could also be handled.

At night users would want Hobbit to be silent and dark. As a consequence, Hobbit should not have any lights on when it is not active or sleeping in the charging station. The LEDs at the

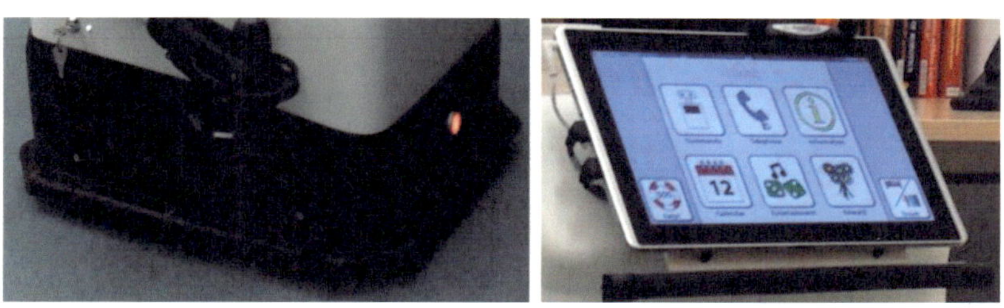

Figure 5. Safety bumper all around the base plate (left, only the front view of the robot is shown) and (right) tablet base bumper (the black rubber band below the tablet).

switches of Hobbit, the touch-screen on the charging station and even the eyes of Hobbit are too bright at night when the user wants to sleep. Similarly, when the user presses the Away/ Sleep button, the robot should not radiate light.

5.5. Considerations for the design of the robot hull

The outside hull to cover all the body is essential for the presentation of the robot. It is important to have a good first impression as well as high functionality. The design carried out in Hobbit turned out to be very good and provided all the features necessary, for example, the drinking bottle mounted low on the robot for emergency cases.

As a start of the investigations into the Hobbit robot, several workshops had been conducted with older adults for indicating the requirements from the side of the users to the robot hull. **Figure 6** shows some of the example robot models created. Often modelled characteristics are soft but washable and hygienic materials, a slim body and two arms. Due to practical reasons and cost issues, only one arm has been realised. For all the functionality integrated in the top part of the hull and head, see the next section.

An issue regarding a safe robot motion is the mounting of the robot arm on the platform. Ideally, the entire arm should be within the limit of the footprint of the robot base platform. This is a challenge for the technical components: mobile arms of small size and reasonable payload do not yet exist. It also needs to be taken into account that during arm motion planning the arm motion will come out of the base footprint and hull. But it needs to be considered for avoiding self-collisions and collisions with the environment. Moving the arm and specifically the shoulder in a few centimetres is actually possible and is feasible in a final robot system concept.

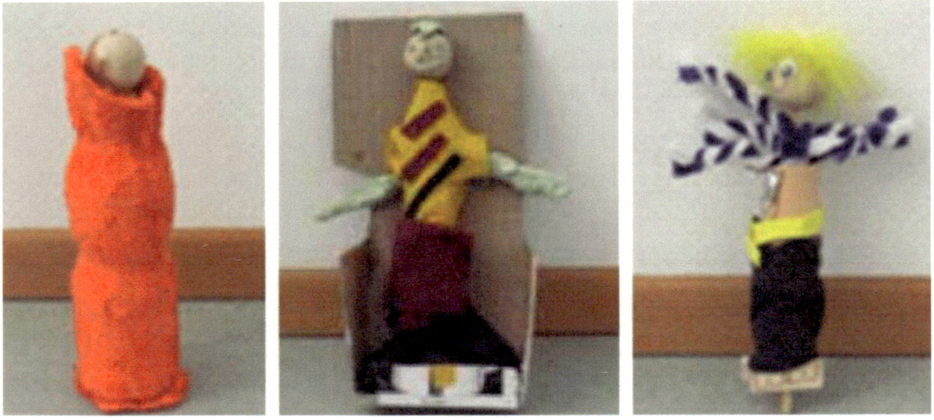

Figure 6. Three of the proposals for the robot design from workshop participants using simple materials to build up a robot model. Note the common characteristics of soft materials, a slim body and two arms (the model of the left robot does not have arms but the description of the user).

The design itself has been very good. A further functional improvement regards the practicability to mount and unmount the hull in cases of changes. This is of particular interest for earlier phases where this needs to be done more often but also for a product to allow good maintainability. For Hobbit we rather had the design and then considered how to best split up the hull for manufacturing as well as practical issues. In the next version, this aspect needs to be included into the design considerations already at design time.

One more issue of practical use, in particular in the phase of testing the robot and when introducing the robot to first home and user trials: data of the platform should be easy to access, for example, via a service socket. This is relatively easy to achieve by adding the necessary sockets, for example, USB to copy data rapidly or attach a keyboard, to add a network if there is a WLAN error or to connect a display to the system if the XPC does not start, or it is needed to change settings in the BIOS. This socket could be included into the design considerations such that it can be easily accessed without removing the hull. This will considerably simplify and speed up the development of the platform behaviour and performance.

Furthermore, the hull should be included in the heating concept, for example, to use materials that provide a better cooling or the use of ventilation slots that were not foreseen in Hobbit because of the risk of liquids by transporting water. In particular during the summer trials, the robot produced too much heat, and this was felt to be negative by the users.

5.6. Multiple functionalities for the active robot head

As the last section of future improvements, we investigate more detail how the active head of the Hobbit robot worked and what could be improved. The active head in itself actually

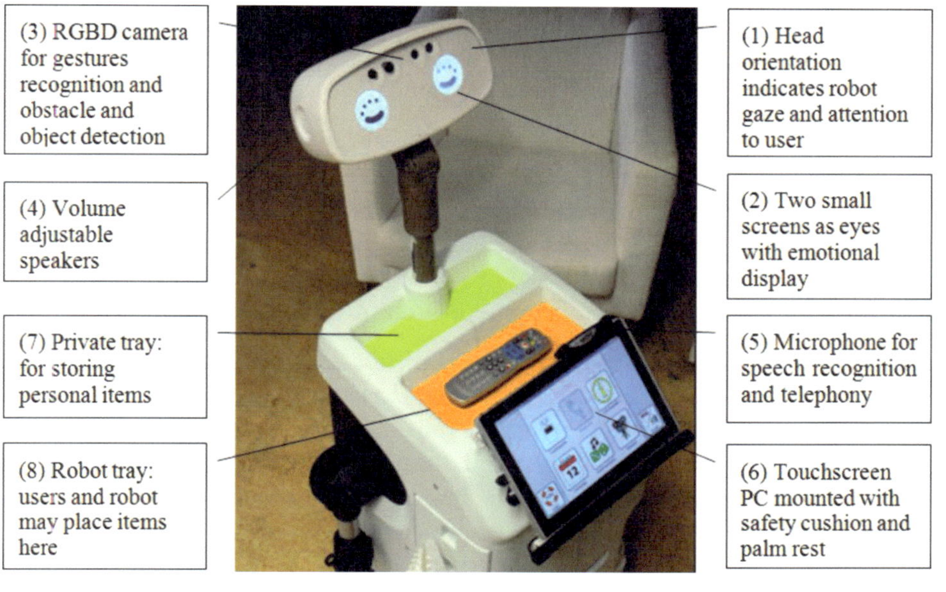

Figure 7. The upper body and the head of Hobbit. Also, depicted are the main components and functionalities.

turned out to be very useful. Note that users had indicated that there should be a head to make clear where to talk to. A main function the head and upper torso of the robot serve is to give the user a feeling of active presence of Hobbit. **Figure 7** gives more details about the upper body and the head. The key function of the active head is to enlarge the field of view of the otherwise limited RGBD sensor. With the head it now covers different functions including the detection of obstacles and humans, the recognition of gestures, the recognition of objects and the detection and manipulation of objects for pick-up from the floor. Blue Danube Robotics (www.bluedanuberobotics.at) designed and produced the active head.

Besides increasing the field of view, the head serves several other important functions. We list the most important ones (see also [6]):

• Attending to the user: an active head naturally presents the direction into which the robot is facing. The older adults in the user trials intuitively understood where the robot is looking. This renders the head motion very efficient: users immediately understood if the robot was busy with navigation when looking straight down or grasping/searching an object when looking around. When approaching the user, the robot would raise the head, and using human torso and face detection focuses the user. No additional means such as lights of verbal communication is necessary.

• Faster search operations: the active head does not need further robot motion to look all around. Moving the entire platform is much more expensive and all measures than simply moving the head. Also, with the up and down motion, near and far views can be sampled. There is the drawback that reliable operation needs a calibration of the pan/tilt motion. However, in the near future, we will present an automatic and continuous calibration function simply from moving the head.

• Detecting obstacles for reliable and safe navigation: the robot looks straight down when navigating. This allowed us to cover in more detail the ground in front of the robot. This is of particular interest when approaching a user, since hitting a foot is definitely not allowed with older adult users. It also covers the part of the blind area in front of the robot from the bottom RGBD camera for detecting walls and localisation. The touch-screen is the limit at present. But one might think of a detachable touch-screen or tablet, which further increases the operational range.

The present field of view (FOV) of typical RGBD sensors is about 58° horizontally and 43° vertically. A FOV with this limitation is presented by more or less all RGBD cameras including the Kinect. Sensors with other measurement principles such as time of flight (TOF), for example, the KinectOne, have about the same limitation. It may be interesting for sensor developers to think about this or to provide a direct integration with an active pan/tilt unit as proposed here.

Nevertheless, the present active head serves the purpose of attending to many different tasks such as viewing the floor in search of objects for possible pick-up, detecting obstacles during navigation, checking tables for objects that the user can ask for, the detection of persons, the recognition of gestures or user activities and the search for the user to increase her perceived feeling of safety. Particularly, in the last task, finding the user, a large FOV is needed. While the human visual FOV covers about 180°, the present head set-up needs three viewing directions to cover the same range. Still, with present camera resolution and the added advantages of an active head as listed above, we see this as the best set-up for a future home robot.

The functionality of an active gaze direction to show the user where the robot looks helps to create a bonding with the user as indicated by the trials. More work in this direction is ongoing, for example, [1, 12]. What is still missing are robots that robustly and smoothly focus onto the user in their natural environment at home. Another interesting point is that there is often a fixed approach to the user. However, at home, the exact location of chairs changes, and hence the location of the sitting user changes. This renders necessarily an adaptive approach to the older adult. In Ref. [13] a laser distance measure to the legs of a single user was used [13]. However, a more flexible and accurate method including the recognition of the user is asked for at-home robotics.

Let us give a few more insights to improve the head and robot operation. At present the faces are partially black and white. A completely dark face better hides the dark holes from the camera, whoever may not look appealing. Hiding the camera behind glasses disrupts camera accuracy and calibration. At present there is not perfect solution yet.

Another useful option is remote accessibility using a secure Internet link. It provides developers to keep track of the robot and, if any problems are encountered, to rapidly check without the need to physically reach the robot test site. Similarly, the hull should be easily detachable. If problems with the robot hardware need to be resolved, this will prove very useful.

An addition that may be necessary, depending on the environment, is ultrasonic sensors (US). In the cases we encountered, this was not necessary, but transparent surface is not visible to optical cameras such as the RGBD sensors and needs a complementary sensor modality such as US. Similarly, at present we did not have any sensors on the robot's backside. This was resolved by not driving backwards except out of the docking station, where it was known that there is space and the user was verbally notified that this would happen. An option is to increase the range of the active head to also look backwards. Another option is to add more US.

Finally, let us conclude with an important finding regarding the robot's safety. From the experiences in Hobbit and also other robot projects at home and in office settings, there are many cases where stairs lead downwards. While looking down at the floor is an option, this is a safety critical aspect, and, hence, it should have at least two if not three complementary measures to assure that the robot is driving forwards on safe ground. Floor detectors that use infrared sensors and blocking areas in the navigation software are too easy to realise options to considerably enhance the robot's operational safety. All needs to be integrated into the navigation system, but it will prove useful. As a manufacturer of a mobile robot clearly said, it will be able to fall down but only once.

6. Conclusion

This chapter presented the newly developed service robot called Hobbit. We built Hobbit with the purpose of assisting older adults and to improve their perceived feeling of safety at home. The ultimate goal is to make users stay longer in their homes by using new information technology and new solutions such as smart environments (including ambient-assisted living (AAL)). Besides fostering the feeling of safety, we also set out to improve the feeling of self-efficacy, that is, one's own ability to complete tasks. We set out to achieve this by providing a rich set of functions to the user. We provided methods for emergency detection including the patrolling to regularly locate the user and start an interaction and the detection of emergency situations as a means to directly react if needed, and we provided proactive functions to avoid falls in the first place. These measures included to keep the floor free of clutter, to transport items so the user can walk with hands free, a set of fitness functions to stay active with cognitive and physical exercises and the option to freely set reminders, also with the idea to stay as active as possible.

After the summary of the questionnaires highlighting the functionality, safety and operation features that older adults would want, we presented the Hobbit platform and its technical components. We then presented details of the results of a longer study of the Hobbit robot in the wild—in the homes of 18 older adults. The study lasted for 3 weeks each. Another specific highlight of the study was that the robot was navigating autonomously, a difference to other present robots operating at home.

Of particular interest in the user trails was that the users could freely select what they wanted to do with the robot. This is a considerable change over other studies, for example, with robots such as Kampai or Care-O-bot, where the sequence of trials is scripted and repeated at a set times. An example is the reminder to drink, which in a single test run may be fun for the user; however, the repeated reminder to drink will be rejected by users. In the free setting, this is different, since users select what they would want. For example, users set the reminders all by themselves or with the help of the person.

In conclusion, the mutual care (MuC) concept, which has been proposed to foster the acceptance and improve the use of robot, has been detected to have rather little consequence. A cause for this unexpected result may be with high probability that the technical functioning of the robot is not yet high enough. Hence, there is considerable work ahead. While the trials indicate a first proof of concept, there needs to be more added reliability of all the robot operations. The good finding is that navigation is autonomous and with a service rate of over 98% was rated as sufficient by users. Hence, we provided a very good start for gaining acceptance by the users. This all indicates that reliability is a prerequisite for user acceptance. It will take next steps to work both on the hardware and the software to reach a higher level of reliability not only for navigation but also the other functionalities. Of particular interest are approaches that cope with automatic calibration of sensors, actuators and the sensor-actor combination. This encompasses the mobile platform and the sensors for navigation as well as the active head and the mobile manipulator. It also includes the reliable detection of drivable floor. As many of these capabilities could be reached with adequate software, it will strongly add to the cost-effective concept provided by the Hobbit platform as presented here.

Another important aspect of Hobbit is the HRI. Considering that it should be effective, our main insights are that the active head proved to be paramount to present information about the robot status and, thus, it facilitates HRI in itself. Furthermore, the active RGBD sensors of the head drastically enlarges the field of view of the otherwise restricted camera, and it serves several robot functions for navigation, search and grasping objects. Actually, the active head was found as the key element of intuitive and easy-to-use HRI: the head viewing direction directly indicates what the robot is doing. Augmented with verbal output, the robot lets the user always know what it is doing, and for the user, it is obvious when she can address the robot or not, because it is busy when looking down. Additionally, the adaptive approach to the user and the direct facing of the user create attention: the user knows that now the robot is ready to be addressed and accept the next command. In summary, it would be good to give active robot heads more attention. Remember, as depicted in **Figure 6**, users wanted a head. It is the obvious part of the robot to speak to. Hence, it largely facilitates HRI. To be more fluent, it should be given more research and development work. Furthermore, increasing head range to even look back may alleviate the need for sensors in the back and further increase safety aspects.

To conclude, the Hobbit study in the wild shows a mixed picture. While certain functionalities such as grasping objects from the floor and searching for objects do not provide sufficient reliability, other functions such as navigation, docking, entertainment, reminders and the fitness function proved to be very useful and welcome by the users. Users saw the great potential of such a robot: the capability to pick up something from the floor and to transport and find objects is very highly evaluated. They would not want a cheaper robot with reduced functionality, e.g. without an arm. It is rather that they would want a few more functions such as searching for objects.

6.1. Future developments

These results clearly give advice for future research directions. The robot hardware itself needs to be drastically reduced in size, rather 40 cm in diameter than 56 cm in the diagonal. The robot height should be slightly reduced to 120 cm. A height has been nearly reached to good effect in Hobbit. The only other robot with this height is Pepper at present. Other robots are higher. The smaller robot size will also make it easier to navigate in the typically tight spaces in the homes of older adults.

Based on the results from the questionnaires and interviews, the reminder functions, picking up objects and bringing objects, are all highly validated by users and thus need to be considered as high-priority user requirements. Picking up objects and bringing objects need an arm and some sort of gripper. Users wanted the gripper and saw, in particular, the advantage of picking up things from the floor.

There are a few other clear misses in the present Hobbit robot. One that most users would want is a speech interface. It clearly needs to be part of Hobbit's user interface. Speech recognition and output therefore need to be state of the art, without ignoring the final objective of the project of affordability of Hobbit. The technical problem at present is distant speech recognition. While speaker into a headset or the mobile phone provides good recognition results, a microphone mounted on the robot, the large distance to the user and the noise of the robot itself are all factors to limit speech recognition drastically. While on the one hand an improvement of distant speech recognition can be expected, another option may be to equip Hobbit with a mobile phone as a natural interface to the robot. While considerations for this solution have been made, the main issue to solve is that the mobile phone could be replaced. Hobbit will need a functionality to search for it using RFID chips and sensor or other means.

Another aspect that is of need for future developments is an adaptive behaviour from Hobbit. Depending on the preferences of the user (i.e. an individual user profile), the amount of autonomous, proactive behaviour of the robot needs to differ. Ways of adapting to the user profile are to be developed for every function from approaching the user, to the times and frequency when the robot should approach the user. Ideally, this is learned from the first trial interactions. Though, this will require substantially more research and improved capabilities of the robot to fully understand the user.

Finally, the Hobbit study with letting the robot be operated with any prescripted operations showed that older adults are indeed very interested in new technology and that a robot at home has very high potential. Although the present price tag may be too high, users clearly indicate that the full functionality is of the highest value and should be pursued in further developments.

Acknowledgements

This research has received funding from the European Community's Seventh Framework Programme under grant agreements No. 288146 Hobbit and No. 610532 Squirrel and funding from the Austrian Science Fund under grant ALOOF No. I1856-N30. Thanks to Andreas Bley of MetraLabs for the picture of the robot docking station.

Author details

Markus Vincze*, Markus Bajones, Markus Suchi, Daniel Wolf, Lara Lammer, Astrid Weiss and David Fischinger

*Address all correspondence to: vincze@acin.tuwien.ac.at

Technische Universität Wien, Vienna, Austria

References

[1] Das D, Rashed MG, Kobayashi Y, Kuno Y. Supporting human-robot interaction based on the level of visual focus of attention. IEEE Transactions on Human-Machine Systems. 2015;**45**(6):664-675

[2] Rebecca E. Grinter Ja-Young Sung, Henrik I. Christensen. Robots in the wild: Understanding long-term use. In: ACM/IEEE Conference on Human Robot Interaction (HRI). 2009

[3] Leite I, Martinho C, Paiva A. Social robots for long-term interaction: A survey. International Journal of Social Robotics. Springer, 2013;**5**(2):291-308

[4] Hanson VL. Age and web access: The next generation. In Proceedings of the 2009 International Cross-Disciplinary Conference on Web Accessibililty, ACM, pp. 7-15, 2009

[5] Astrid Weiss, J. Beer, T. Shibata, and M. Vincze. Socially assistive robots for the aging population: Are we trapped in stereotypes? In Proceedings of the 9th ACM/IEEE International Conference on Human-Robot Interaction, 2014

[6] Markus Vincze, David Fischinger, Markus Bajones, Daniel Wolf, Markus Suchi, Lara Lammer, Astrid Weiss, Jürgen Pripfl, Tobias Körtner, and Christoph Gisinger. What older adults would like a robot to do in their homes - first results from a user study in the homes of users. In: IEEE International Symposium on Robotics ISR, 2016

[7] Lammer L, Huber A, Weiss A, Vincze M. Mutual care: How older adults react when they should help their care robot. In: . Proceedings of the 50th Annual Convention of the AISB; 2014

[8] Pripfl J, Körtner T, Batko-Klein D, Hebesberger D, Weninger M, Gisinger C, Frennert S, Eftring H, Antona M, Adami I, Weiss A, Bajones M, Vincze M. Results of a real world trial with a mobile social service robot for older adults. ACM Digital Library. 2015

[9] Fischinger D, Einramhof P, Papoutsakis K, Wohlkinger W, Mayer P, Panek P, Hofmann S, Körtner T, Weiss A, Argyros A, Vincze M. Hobbit, a care robot supporting independent living at home: First prototype and lessons learned. Robotics and Autonomous Systems. 2016;**75**(Part A):60-78

[10] Yardley L, Beyer N, Hauer K, Kempen G, Piot-Ziegler C, Todd C. Development and initial validation of the falls efficacy scale-international (fes-i). Age and Ageing. 2005;**34**:614-619

[11] Huber A, Lammer L, Weiss A, Vincze M. Designing adaptive roles for socially assistive robots: A new method to reduce technological determ stereotypes. Publication in the International Journal on HRI. 2014;**3**(2):100-115

[12] B. Pierce, T. Kuratate, C. Vogl, and G Cheng. Mask-bot 2i: An active customisable robotic head with interchangeable face. In: IEEE-RAS International Conference on Humanoid Robots (Humanoids), 2012

[13] Ross Mead and Maja Mataric. Robots have needs too: People adapt their proxemic preferences to improve autonomous robot recognition of human social signals. In: New Frontiers in HRI Symposium, AISB, 2015

Chapter 3

Evaluation and Fault Classification for Service Robot during Sit-to-Stand Movement through Center of Mass

Tianyi Wang, Hieyong Jeong and Yuko Ohno

Additional information is available at the end of the chapter

http://dx.doi.org/10.5772/intechopen.72176

Abstract

Many service robots have been developed to assist patients with sit-to-stand movement (STS). However, little research has focused on users' negative psychological changes during the STS movement when assisted by a robot. The STS movement accompanied with a negative psychological change is defined as a fault. The main purpose of this study was to propose a method of conveying faults to a service robot through the center of mass (CoM). Experiments on the STS movement were executed five times with 10 healthy subjects under four conditions: two self-performed STSs with seat heights of 43 and 62 cm, and two robot-assisted STSs with a seat height of 43 cm and end-effector speeds of 2 and 5 s. Time series data on the CoM were measured with high-speed camera system. A classifier was designed according to the data on the CoM in the frequency domain. The results showed that the proposed classifier had a high probability of discriminating fault classes from others. Then, the vertical ground reaction force (vGRF) under the same experimental conditions was used to cross-check the experimental results. It was concluded that faults in the assistance of service robots can be detected from the CoP-related items.

Keywords: center of mass, fault classification, service robot, sit-to-stand movement

1. Introduction

The sit-to-stand (STS) movement is essential to daily life and is used to change from a sitting position to a standing position. Dall and Kerr [1] pointed out that healthy adults performed 60 ± 22 STS movements every day on average. Grant et al. [2] reported that healthy older adults living in the community performed significantly more STS movements per day (71 ± 25) than older adults attending a day hospital (57 ± 23) or frail older patients in a rehabilitation ward (36 ± 16). For the elderly, their ability to perform this basic task weakens from the deterioration of muscle strength, joint range of motion, and balance. This can increase the risk of institutionalization, impaired functioning and mobility for activities of daily living (ADL), or even death [3–5]. With the continuous decrease in the number of nursing care specialists, there is an urgent need for the development of STS-assisting service robots.

Only a few studies have focused on service robots for assisting STS movement. Gervand et al. [6] formulated unassisted and assisted STS transfers as a control problem and found a balance among the end-point accuracy, human balance, energy consumption, and smoothness of motion. Chuy et al. [7] presented two approaches to assisting with the STS movement by using a robotic walking support system and showed that this system can track the desired support force. Rather than using kinematic data, Burnfield et al. [8] compared muscle demands through electromyography during self-performed and device-assisted STS transfers.

However, studies on service robots have neglected the psychological impact on subjects so far. A good understanding of psychological issues may help facilitate better integration of medical devices with users. Thomson et al. [9] emphasized that medical devices have both positive and negative psychological impacts on users, and simply addressing safety requirements does not guarantee user satisfaction. Because the duty of a service robot is to provide satisfaction with its service, cases where robot fails to meet this need can reasonably be defined as a fault.

A robot does not have the ability to tell whether its service satisfied its users. In other words, present service robot is not aware of its fault. Inferring human psychology is an essential step toward understanding human actions and hence is critical for realizing human-robot interaction. Recent advances in sensors and algorithms have allowed researchers to improve the perception ability of robots. However, only perception may not be sufficient for the efficient interaction between a human and robot because the robot's reaction should depend on its understanding of human actions. This observation raises the question of how to correlate human psychological changes to legible data. Answering this question can provide methods for giving feedback to a service robot so that it can offer better health care service and daily-life assistance. In contrast to a laboratory environment, the situation in hospitals, rehabilitation centers, and nursing homes always have conditions that do not allow the use of large-size experimental equipment such as motion capture system or traditional force plate board. Thus, our aim in this study is to propose a practical, economical, and reliable method of evaluating service robot and conveying faults to service robot. Psychological results were collected by asking subjects to answer a questionnaire. The results showed that the proposed method discriminates faults with a high probability and is suitable for clinical research.

This chapter is organized into five sections. Section 2 explains the subjects, experimental system, and proposed method for analyzing the STS movement under different experimental conditions. Section 3 describes the experimental results. Section 4 cross-checks the results with center of pressure (CoP) and introduces other advanced signal analysis and classification approaches. Finally, Section 5 concludes this chapter.

2. Subjects and methods

2.1. Subjects

Most service robots were originally designed for the elderly or patients with particular diseases. The elderly and patients were allowed to freely participate in our experiment because no clinical evidence had demonstrated the effectiveness and availability of service robots.

Ten human subjects (age: 38.6 ± 12.2 years old, height: 1.72 ± 0.06 m, body mass index: 22.37 ± 2.60 kg/m^2) volunteered for the experiment. No subjects reported a major back pain or lower limb pathology, use of medication, or history of neurological disease that may influence the standing balance. There was no large difference in BMI for all subjects ($p > 0.005$).

The experimental procedures of the present study were in accordance with the Declaration of Helsinki and approved by the Ethics Committee on the Division of Health Science, Graduated School of Medicine, Osaka University (No. 305, August 21th, 2014). Informed consent was obtained from all subjects.

2.2. Experiment system

Figure 1 shows the overview of experiment system. Service robot evaluates experiment system consists of three equipment: a service robot for assisting with the STS movement, a height-adjustable chair for initializing the STS movement, and a high-speed camera system for measuring the STS movement.

Figure 2(a) shows a prototype of service robot whose end-effector speed can be regulated between 1.5 and 5.0 s. This robot can help users stand up, walk, and sit down. **Figure 2(b)** shows each step of the STS movement with assistance from service robot. The steps followed the definitions provided by Schenkman et al. [10] and are marked by five events. The path of

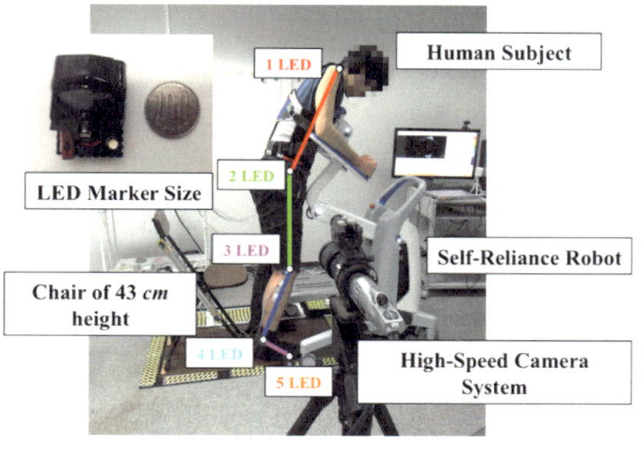

Figure 1. Experiment system. The high-speed camera system consists of a high-speed camera (250 Hz, VW-9000© KEYENCE, Japan) and five LED markers placed on the users at the following position: acromial, greater trochanter, lateral patellar ligament, lateral malleolus, and instep. Each marker was placed according to the ISO 7520: 1996 standard. Each maker is nearly the size of, but much lighter than a 100-yen coin.

(a) prototype of robot for STS support

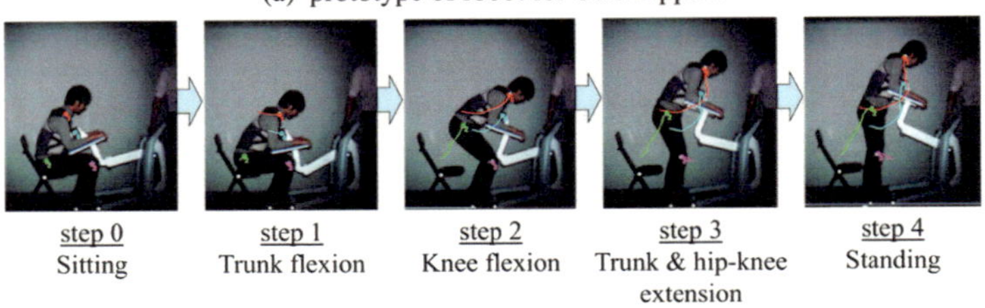

(b) steps of STS movement with support of robot.

Figure 2. Overview of service robot. (a) A prototype service robot, complying with the ISO/TS 15066:2015 standard, with end-effect forces of 240 and 100 N in horizontal and vertical directions, respectively; (b) steps in robot-assisted STS movement, step (0) sitting still; step (1) trunk flexion; step (2) knee flexion; step (3) trunk and hip-knee extension; step (4) stabilization.

the end-effector could be controlled so that service robot could match the user's trunk movement with self-performed STS. Individual differences in body height could be also accounted for through the control of service robot.

Figure 3 illustrates a design scheme and trajectories of service robot utilized in our study, which is a robot entailing a chain-motor system containing a geared motor and three chains, see **Figure 3(a)**. Trajectories of robot arms and end-effector are presented in **Figure 3(b)**, wherein the start and end points are programed to match the position of user's chests while they are sitting and standing still. At present, this service robot can only support users' trunks.

A chair (CS-320A) was used to adjust the seat height during STS movement. The model could adjust the height to six levels (32–62 cm). The chair had the dimensions of a 35 cm width, 48 cm depth, and 78 cm height.

In order to acquire the details of users' STS movement, we used a high-speed camera system, where there is a high-speed camera (VW-9000©KEYENCE, Japan) with 250 Hz sampling frequency and horizontal and vertical resolution of 0.5 mm (set through pre-experiment calibration). Five LED markers were placed on the users at: acromial, greater trochanter, lateral patellar ligament, lateral malleolus, and instep. Each marker was placed according to the ISO 7520:1996 standard.

http://dx.doi.org/10.5772/intechopen.72176

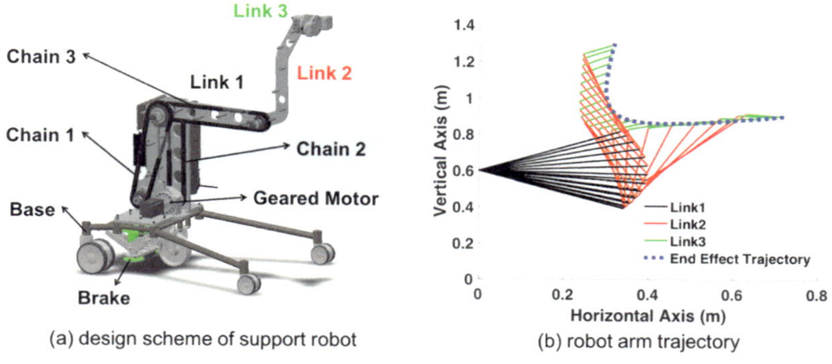

Figure 3. A 3-D design scheme of service robot and trajectories of robot arms and end-effector utilized in this study.

2.3. Class design

Janssen et al. [11] categorized determinants that may result in stress during the STS movement such as chair-related factors. The literature indicated that the chair influences the performance of STS movement. Lowering the height of the chair apparently makes the STS movement more demanding or even unsuccessful. For young subjects (25–36 years old) without impairments, lowering the seat of chair from 115 to 65% of the knee height increased the trunk flexion angular velocity to stand by almost 100%. Accordingly, seat heights of 62 cm (>115% of subject's knee height) and 43 cm (<115% of subject's knee height) were set as Classes 1 and 2, respectively.

Although various determinants influence the STS performance, once the STS movement is assisted by the service robot, most determinants including the age, gender, foot position, trunk position, and arm movement are restricted by the robot. The assisting speed of the service robot naturally becomes the main factor that may influence how the STS movement is performed and directly affects the psychological changes of the user. Thus, we set up two robot assistance classes: Classes 3 and 4 correspond to end-effector speed of 2 and 5 s, respectively, for a chair height of 43 cm.

In order to ensure that subjects got used to the experimental procedures and assistance from the service robot, all subjects performed several training runs before the experiment. Hesse et al. [12] emphasized that training can be a determinant in an experiment study. For the self-performed STS experiment, subjects were told to stand up from the chair with their self-selected speed five times. For the robot-assisted STS experiment, because the trajectory of the end-effector was fixed and subjects had different body heights, it was difficult to make the robot accommodate every subject. As a result, robot-assisted STS was only performed with subjects at a suitable height five times in the experiment and utilized for analysis.

As a medium between psychological changes and physical movement, subjects were asked to answer a questionnaire on aspect of the STS movement that they felt demanding during both self-performed and robot-assisted experiments. Based on the results of the questionnaire, we were able to define a particular class as a fault when accompanied by a demanding feeling.

2.4. Fault measurement

The center of mass (CoM) and center of pressure (CoP) of the human body are widely utilized to analyze human motion. The location of CoP is highly variable, especially when vertical forces are small near the beginning or end of the stance phase. In addition, the trajectory of the CoP only represents the location of the vertical ground reaction force vector. It is difficult to understand the mechanism and interaction between the upper and lower limbs. Consequently, a method that can vividly describe human motion is necessary.

CoM is frequently calculated and proven to be significant especially in studies on human postures. As the location of CoM is not fixed and changes continuously with alteration of postures, the CoM needs to be calculated with time series data, and details about the CoM trajectories of each body segment are required.

Figure 4 shows an example on how to calculate CoM. Center of mass on a segment is located at the point that creates the same net gravitational moment of forces about any point of a segment as did the original distributed mass. We can simplify the CoM of segment as a single mass M located at a distance X_M from end of the segment as:

$$M X_M = \sum_{i=1}^{n} m_i x_i \tag{1}$$

In our experiment, human model was simplified as a three-segment model as shown in **Figure 4**. First, we assume that CoM for a total link model is located at point M_0, coordinate (X_{M0}, Y_{M0}). Coordinates can be calculated as follows:

$$\text{Coordinates} = \begin{cases} X_{M0} = \dfrac{M_1 X_{M1} + M_2 X_{M2} + M_3 X_{M3}}{M} \\ Y_{M0} = \dfrac{M_1 Y_{M1} + M_2 Y_{M2} + M_3 Y_{M3}}{M} \end{cases} \tag{2}$$

Frequently, the mass of each segment M_0 is calculated as a particular percentage of whole body weight. Then we have the calculation of coordinates of CoM of total segment model in our study:

$$\text{Coordinates} = \begin{cases} X_{M0} = f_1 X_{M1} + f_2 X_{M2} + f_3 X_{M3} \\ Y_{M0} = f_1 Y_{M1} + f_2 Y_{M2} + f_3 Y_{M3} \end{cases} \tag{3}$$

where X_{Mi} and Y_{Mi} are the horizontal and vertical coordinates of CoM of each segment.

The abovementioned questions make the calculation of CoM simple because the only knowledge required is the fraction of total body weight and the coordinates of each segment CoM. In this study, f_1, f_2, and f_3 are set as 0.433, 0.433, and 0.5 for shank, thigh, and trunk, respectively.

A time-varying signal can be represented by successively adding individual frequencies present in the signal. Note that the frequency spectrum is mandatory during signal analysis. Thus, we used fast Fourier transform (FFT) to replace the time domain data with frequency domain

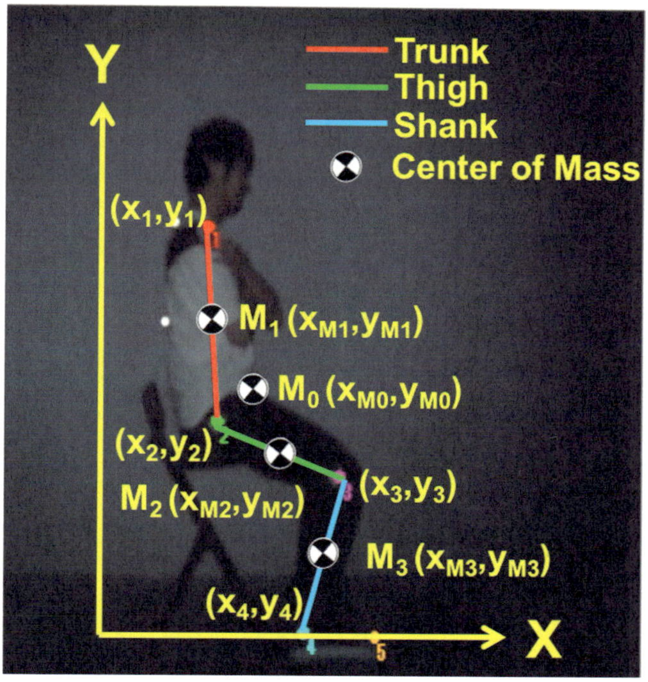

Figure 4. Example of calculation of CoM.

data. Based on the FFT results, we are able to remove undesirable noise and depict all phases during the STS movement without any omissions. Because it is difficult for human to execute the STS movement as fast as 3 Hz, frequency domain data of less than 3 Hz were used in this study. Then, the maximum amplitude and its frequency were utilized as classifying features. We chose the Bayesian classifier for fault classification.

The main indices about FFT are given as follows:

- Feature 1: maximum amplitude value of horizontal velocity of CoM
- Feature 2: frequency of maximum amplitude value of horizontal velocity of CoM
- Feature 3: area of frequency domain data of horizontal velocity of CoM with 3 Hz
- Feature 4: maximum amplitude value of vertical velocity of CoM
- Feature 5: frequency of maximum amplitude value of vertical velocity of CoM
- Feature 6: area of frequency domain data of vertical velocity of CoM with 3 Hz

In order to discriminate features in each class, FFT was calculated from differential data of horizontal and vertical coordinates of CoM. The area of FFT curves was also calculated in order to consider the second and third highest frequency as well as the first highest frequency.

2.5. Fault classification

Bayesian classification and decision-making are based on probability theory and the principle of choosing the most probable or the lowest risk option. The major problem with the Bayesian classifier is the class-conditional probability density function, which describes the distribution of feature vectors in the feature space of a particular class. The distribution can be estimated from a training set with a range of methods. The probability density function is defined as a weighted sum of Gaussians:

$$p(x; \theta) = \sum_{C=1}^{C} \alpha_C N(x; \mu_c, \Sigma_c) \quad (4)$$

where α_c is the weight of the component c, $0 < \alpha_c < 1$ for all components, and $\sum_{c=1}^{c} \alpha_c = 1$. The following parameter list defines a particular Gaussian mixture probability density function: $\theta = \{\alpha_1, \mu_1, \sigma_1, \ldots \alpha_c, \mu_c, \sigma_c\}$.

In D-dimensional space, it is defined in matrix form as:

$$N(x; \mu, \Sigma) = \frac{1}{2\pi^{D/2}|\Sigma|^{1/2}} exp\left[-\frac{1}{2}(x-\mu)^T \Sigma^{-1}(x-\mu)\right] \quad (5)$$

where μ is the mean vector and Σ is the covariance matrix. In order to calculate the discriminant function, Eq. (4) is rewritten as a nature logarithm:

$$\ln p(x; \theta) = -\frac{1}{2}(x-\mu)^T \Sigma^{-1}(x-\mu) - \frac{1}{2} ln|\Sigma| + \text{const} \quad (6)$$

When $\Sigma_i = \Sigma_j = \sigma^2 I$ (unitary covariance matrix), the discriminant function is determined by:

$$y(x) = argmin_i\{(x-\mu_i)^T (x-\mu_i)\} \quad (7)$$

When $\Sigma_i = \Sigma_j \neq \sigma^2 I$ (common covariance matrix), the discriminant function is determined by:

$$y(x) = argmin_i\{(x-\mu_i)^T \Sigma^{-1} (x-\mu_i)\} \quad (8)$$

When $\Sigma_i \neq \Sigma_j$ (general covariance matrix), the discriminant function is determined by:

$$y(x) = argmin_i\{(x-\mu_i)^T \Sigma^{-1} (x-\mu_i) + ln|\Sigma|\} \quad (9)$$

The boundary condition can be determined by using Eqs. (7)–(9).

3. Results

3.1. Results of fault classes

Based on the psychological changes, for self-performed STS, all subjects felt a 43 cm chair height (Class 2) was demanding. For the robot-assisted STS movement, adjusting the assisting speed affected the psychological change. Specifically, all subjects reported that they felt supported with an assisting speed of 2 s. However, when the assisting speed was slowed to 5 s, executing the STS movement became demanding. Accordingly, we defined Classes 2 and 4 as faults.

3.2. Results of calculated CoM and faults classification

Figure 5 shows the experiment images and results of CoM at time domain and frequency domain. **Figure 5(a)** shows four-frame multiple exposure photography to describe subjects' performance under four conditions. **Figure 5 (b)–(d)** shows time series data of CoM at time domain, including trajectories of CoM, CoM velocities at horizontal (X) and vertical (Y) directions. **Figure 5(e)** and **(f)** shows results of CoM data at frequency domain with FFT.

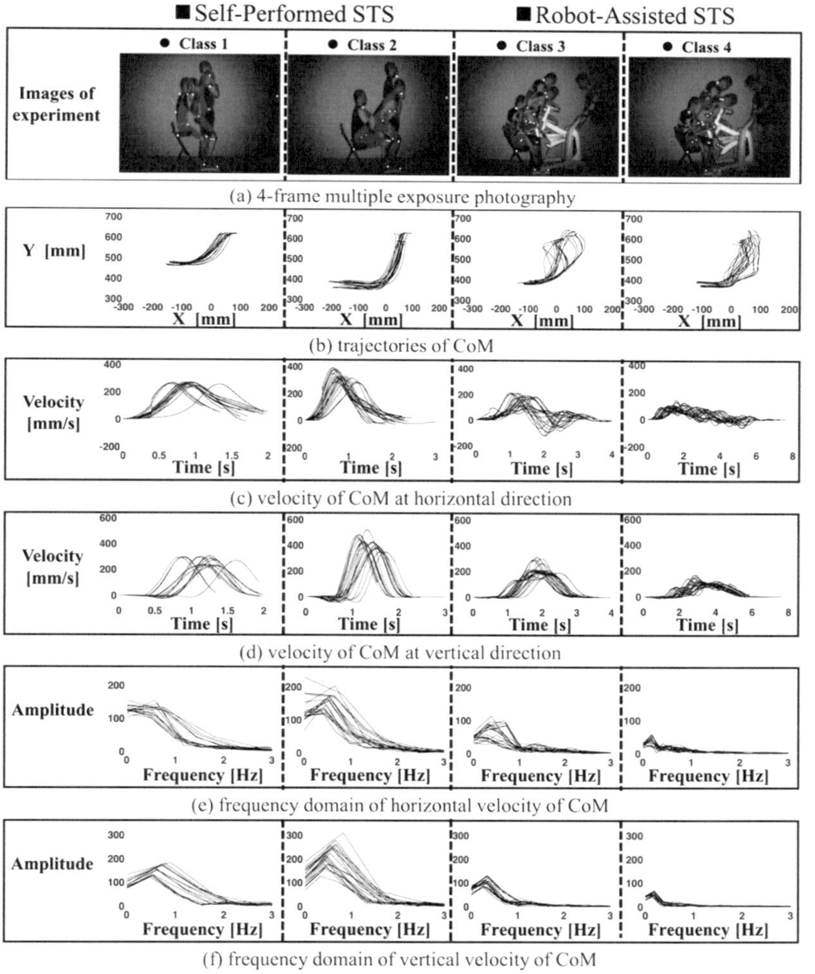

Figure 5. Experiment images, results of CoM, and its frequency domain data. (a) Four-frame multiple exposure photography of experiment; (b) trajectories of CoM; (c) velocity of CoM at horizontal direction; (d) velocity of CoM at vertical direction; (e) frequency domain data of horizontal velocity of CoM with FFT; and (f) frequency domain data of vertical velocity of CoM with FFT.

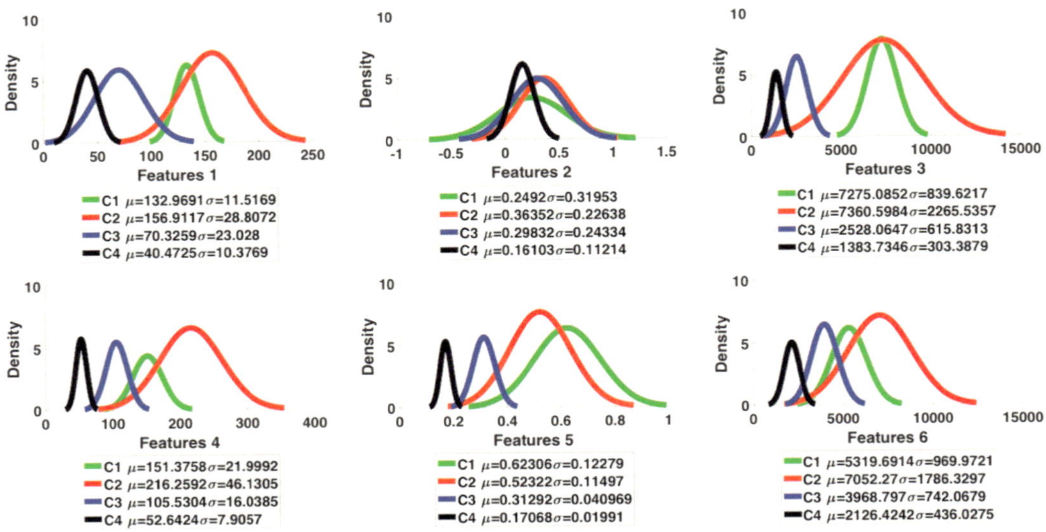

Figure 6. Probability density of six features.

It is obvious to see that due to different seat heights among experimental conditions, CoM trajectories under Class 1 were lower than those under other classes. However, much shorter trajectories of CoM were found under Class 1. As to the velocities at X and Y directions, lowering the seat height increased the CoM speed at both directions almost two times at the

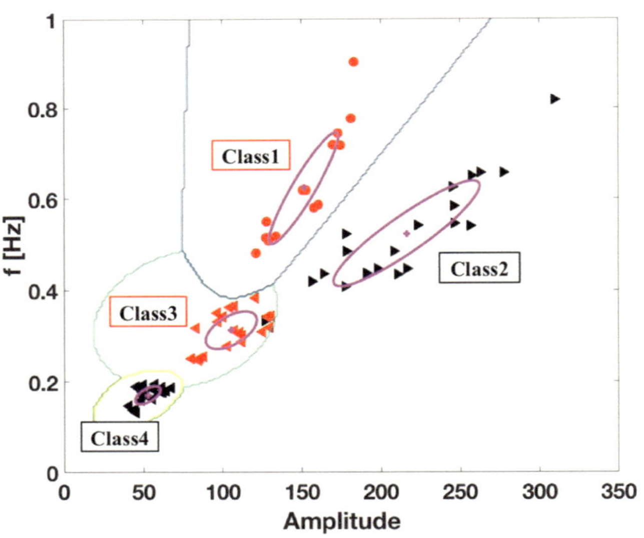

Figure 7. Faults classification results. The horizontal axis represents the maximum amplitude of the CoM in the vertical direction, and the vertical axis represents its frequency. Red circles represent class 1, black right triangles represent class 2, red left triangles represent class 3, and black left triangles represent class 4.

horizontal direction and 1.5 time at the vertical direction. On the other hand, velocities under Class 4 were smaller than those under Class 3 due to slower assisting speed.

From the raw data of frequency domain, it is found that self-performed STS movements showed a dramatic difference compared with the robot-assisted STS movement. It is necessary to identify which indices hold the most probability to distinguish four classes in better way.

Figure 6 shows the results of probability density analysis (PDA) for six features, which can be calculated from **Figure 5(e)** and **(f)**. The major contribution of PDA is to demonstrate in which feature factor, the proposed classification approach would distinguish different classes much better. Principle on how to choose the appropriate features is to find out the largest disparity of distributions.

Accordingly, the maximum amplitude of CoM velocity at the vertical direction (feature 4) and its frequency (feature 5) were used to further classify process.

Figure 7 shows the classification results. Different classes were divided into four areas. Classes where subjects felt supported were located in the middle of the scatter diagram, and classes where subjects felt demanded upon were located on two flanks. Class 4 was isolated from the other classes and more concentrated; the frequency and amplitude in Class 4 showed the smallest magnitude compared with other classes.

4. Discussion

In our study, senior citizens were free to participate in our experiment, even though many service robots were originally designed for the elderly or patients with functional limitations. Schenkman et al. [13] argued that a more successful approach would be if we understood how healthy individuals execute the STS movement under different conditions and utilize this information to interpret the performance of those with functional limitation or impairments.

Two main aspects were discussed in this chapter. First, we cross-check our experiment results through CoP using the vertical ground reaction force (vGRF). Second, we reviewed works on the advanced signal analysis method and other promising fault classification approaches.

4.1. Cross-check of experiment results

Compared with CoM, which requires the masses of body segments to be known and is often not directly determined, the CoP position is the projection on the ground plane of the centroid of the vertical force distribution; thus it can be obtained directly from a force plate or Wii Balance Board (WBB).

The WBB (23 cm × 43 cm platform), which was designed to support people weighing up to 136 kg, fed data into the computer through a Bluetooth connection. Soangra and Lockhart [14] reported that the force sensors are linear with CoP noise levels of approximately ±0.5 mm. Although it was originally designed as a video game controller, the WBB has become a proven tool for assessing the CoP position and has been confirmed to be both accurate and reliable.

Moreover, the WBB provides a portable and inexpensive balance assessment system that is widely available. Jeong et al. [15] used a WBB to analyze the difference in manual material handling between experts and rookies at a logistics workplace. Huurnik et al. [16], Clark et al. [17], and Park and Lee [18] demonstrated the validity and test–retest reliability of the WBB by measuring the CoP position and comparing the data with that from an identical study using a laboratory-grade force plate. They found that the WBB can provide reliable and consistently repeatable data. Yamada et al. [19] reported that vGRF parameters have high reliability with an intra-class correlation coefficient (ICC) of 0.70–0.95. Then, vGRF data on the knee-hip joint extension phase and hip-lift off phase during the STS movement phase can be used to evaluate the lower limb muscle function of the elderly.

Figure 8 describes the WBB and proposed a method for fault measurement. Time series data of vGRF were measured according to the sum of four pressure sensors: P_{V1}, P_{V2}, P_{V3}, and P_{V4}. Because subjects had different body weights, we needed to compare the time series data of vGRF (W'_B) with the body weight W_B. We define this value as $f_{WR}(t)$ and calculated as follows:

$$f_{WR}(t) = \frac{W'_B}{W_B} = \frac{\sum_{i=1}^{4} p_{vi}(t)}{W_B} \tag{10}$$

When subjects sat still on the chair, the vGRF was much smaller than the body weight. It increased and reached its peak value when the trunk and ankle began to extend. Finally, when subjects finished their STS movement and stood still, $f_{WR}(t)$ stayed at a value of unity ($W'_B = W_B$). These measurements are similar to the results of a related conventional study by Yamada and Demura [19].

In order to describe how $f_{WR}(t)$ changed during the STS movement, we defined the differential value of $f_{WR}(t)$ as the STS smoothness. This can be calculated as follows:

$$f_{SM} = \frac{d(f_{WR(t)})}{dt} \tag{11}$$

Figure 9 shows the results about frequency domain data of STS smoothness. In frequency domain, Classes 1 and 2 showed two means and covariance matrices of smoothness (maximum amplitude and its frequency):

$$\begin{aligned} \mu_1 &= [1.1487 \;\; 0.4234], \;\; \Sigma_1 = \begin{bmatrix} 0.0575 & -0.0077 \\ -0.0077 & 0.0030 \end{bmatrix} \\ \mu_2 &= [1.6767 \;\; 0.3861], \;\; \Sigma_2 = \begin{bmatrix} 0.0622 & -0.0060 \\ -0.0060 & 0.0031 \end{bmatrix} \end{aligned} \tag{12}$$

Although there was no large difference between the frequencies of the maximum amplitude for the two classes (0.4234 Hz for Class 1 and 0.3861 Hz for Class 2), the maximum amplitude for Class 2 was 46.0% larger than that for Class 1.

For the robot-assisted STS movement, the maximum amplitude and its frequency for Classes 3 and 4 showed two means and covariance matrices:

http://dx.doi.org/10.5772/intechopen.72176

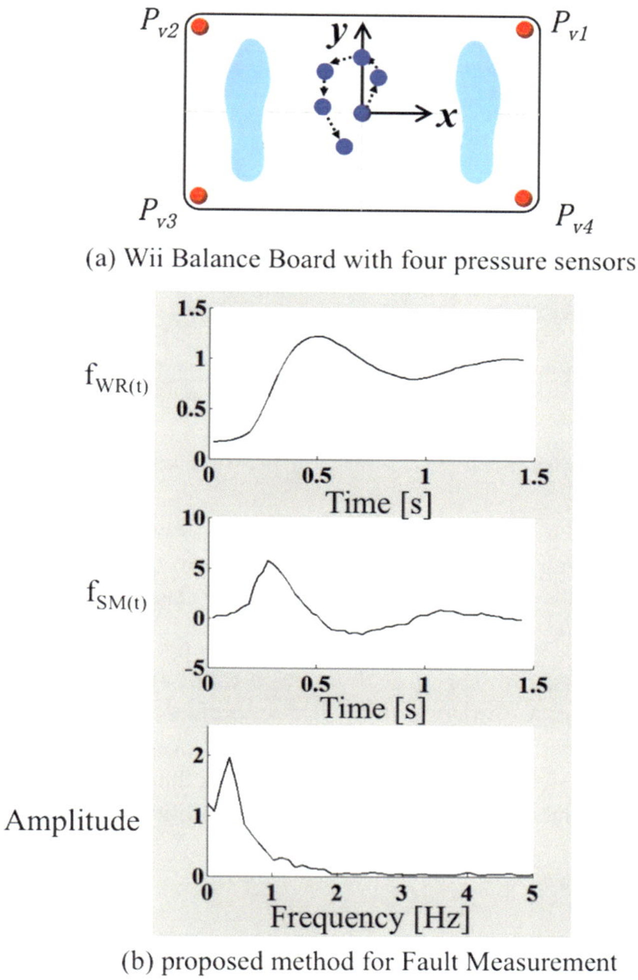

Figure 8. Wii Balance Board and methods for fault measurement.

$$\begin{aligned} \mu_3 &= [0.7697\ \ 0.3659], \quad \Sigma_3 = \begin{bmatrix} 0.0133 & 0.0001 \\ 0.0001 & 0.0033 \end{bmatrix} \\ \mu_4 &= [0.3357\ \ 0.5264], \quad \Sigma_4 = \begin{bmatrix} 0.0007 & 0.0005 \\ 0.00050 & 0.0399 \end{bmatrix} \end{aligned} \tag{13}$$

Note that the maximum amplitude for Class 4 was only 44% of that for Class 3. However, the frequency of the maximum amplitude for Class 4 was 1.4 times larger than that for Class 3.

Figure 10 shows the classification results under four different experimental conditions using same algorithm for CoM. Classes 1 and 2 clearly had different distributions. Even though the subjects were assisted by service robot during STS movement, the psychological results showed that subjects did not feel supported in Class 4. This meant that Class 4 was unique

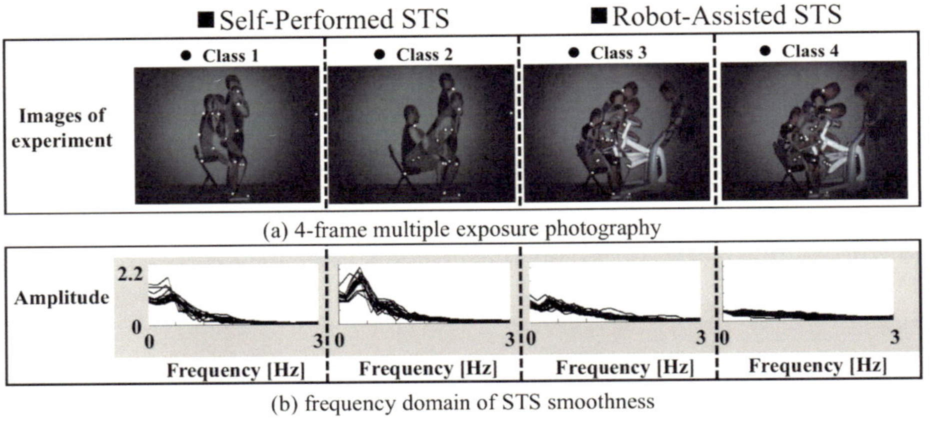

Figure 9. Results of frequency domain data of STS smoothness with Wii Balance Board.

among both the robot-assisted and self-performed STS movement. This phenomenon indicated that nondemanding classes (Classes 1 and 3) were concentrated, while demanding classes (Classes 2 and 4) were located outside the former classes.

After fault classes were classified and discriminated from the others, it was necessary to identify how altering the assisting speed of the service robot affects the STS performance and makes it feel demanding. A difference in the STS smoothness clearly caused a psychological change. Because we considered Class 1 to be natural and normal STS movement, we normalized the STS movement reproducibility through dividing the standard deviation (SD) of peak values of time domain data under four experimental conditions by those for Class 1.

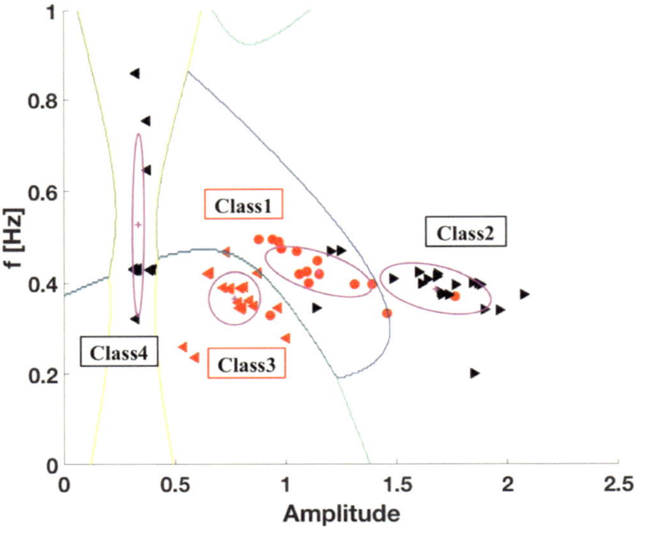

Figure 10. Faults classification results with WBB. Horizontal axis represents the maximum amplitude of STS smoothness and vertical axis represents its frequency. Red circles are Class 1, black right triangles are Class 2, red left triangles are Class 3, and black left triangles are Class 4.

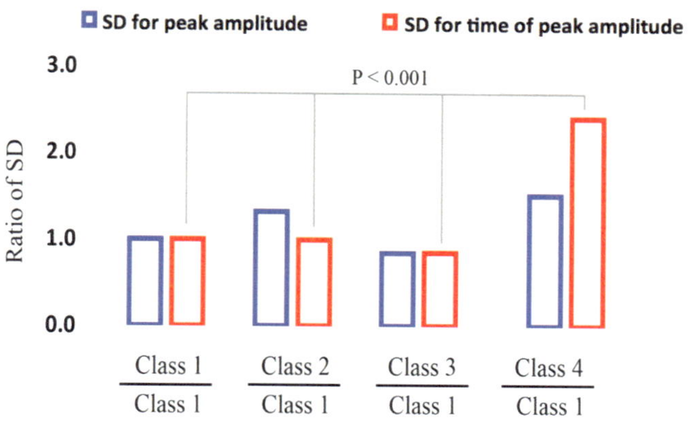

Figure 11. Results of STS reproducibility. The horizontal axis represents each class, and the vertical axis represents the ratio of the SD in different classes to the SD in Class 1. There was no significant difference regarding the SD for the time of peak amplitude ($p < 0.001$).

Figure 11 shows the STS movement reproducibility. Regarding the SD ratio at the peak amplitude, there was no big difference between Classes 1 and 3. However, the value for Class 4 was 1.5 times larger than that for Class 1. Although the trunk flexion and extension were restricted by the end-effector of the robot, the SD for the peak amplitude of the smoothness changed according to the results for Classes 3 and 4. However, the larger SDs for both the amplitude and time were from the effect of lower limbs. Because of the slower speed, it took a longer time to move from trunk flexion (step 2) to trunk extension (step 3). Thus, it was necessary for the lower limbs to produce a larger force to support the upper limbs.

Regarding the SD ratio for the time of the peak amplitude, although there was no big difference between Classes 2 and 3, the value of Class 4 was 2.4 times larger than that for Class 1. A larger SD for the time indicates a different STS movement under the same experimental condition. With Classes 2 and 3, most subjects showed a similar STS movement, but with Class 4, every subject showed different STS movements.

4.2. Advanced signal analysis and classification approaches

In our study, we used a simple questionnaire to describe psychological changes, where negative psychological changes were labeled with different degrees. Subjective emotion measurement also includes interviews on emotional experiences and self-reporting. One of the most widely accepted measures for emotional states was proposed by Russell [20]. The Russell theory consists of a 2-D emotion space defined by the valence and arousal. However, it is difficult for robots to recognize subjective descriptions such as excited or tired. Converting an emotional state to numerical values urgently needs to be addressed.

Swangnetr and Kaber [21] offered an effective approach of inspecting the heart rate (HR) and galvanic skin response (GSR) to measure psychological changes in human-robot interaction through a wavelet analysis. They proposed neural network structures for classification and reported an accuracy rate of about 80%.

In the past 40 years, academic interest in chaos has rapidly increased in several areas, including robotics [22]. Perc expanded the scope of practical application of a deterministic chaotic system to the cardiorespiratory [23] and human locomotor systems [24]. In the former work, basic methods of nonlinear time series analysis were demonstrated to be appropriate for the human electrocardiogram (ECG), and the maximal Lyapunov exponent was calculated to provide a new insight into human heart. The latter study presented an advanced signal analysis that can give overall insight into the dynamics of human motion.

Accordingly, it is reasonable to believe that there is a tight relationship between chaotic theory and machine learning, and it has already been discussed. Khelifa and Boukabou [25] used numerical simulations to confirm the capability of an adaptive neural network to approximate a multiscroll chaotic system. In the future, we intend to draw on the experience of the abovementioned signal analysis method and chaos-based technologies to upgrade our experimental system and fault classification algorithm in order to provide more precise and real-time feedback for service robot.

Besides human locomotor and physiological systems, muscle activation also plays an essential role in our daily life, including the STS movement. Roldan-Jimenez et al. [26] analyzed muscular activity and fatigue using electromyography (EMG) during STS tests and demonstrated that the vastus medialis of the quadriceps (QS) plays a major role in the STS movement. Meanwhile, the QM is also the muscle most likely to be fatigued. It is followed in importance by the tibialis anterior (TA), which was reported to be the second muscle with a high level of participation in STS involvement.

The EMG signal also provides data describing neuromuscular activities and has been investigated as a machine learning tool. Yousefi and Hamiltion-Wright [27] reviewed some classification methodologies using EMG characterizations and concluded that the artificial neural network (ANN) is the most popular method for classifying EMG signals and can achieve a high accuracy of about 97.6%. The second most popular method for classification is the support vector machine (SVM), which has an average accuracy rate of 93%.

However, to the best of our knowledge, there has been no similar work focusing on muscle activities and fault classification for EMG when the elderly stands up from a chair with assistance from a service robot. Thus, as future work, we also plan to explore this area and clarify the relationship between the neuromuscular system and robot-assisting system.

Regarding the limitations of our study, using frequency domain data for a time series signal may be too simple for the undergraduate level. The Lyapunov exponent has been reported to be an excellent approach for analyzing the chaotic behavior of time series data [28]. However, the only concern with using a high-dimensional system for classification is the relatively slow calculation speed, which can directly affect the real-time feedback for robots. In clinical situations, the most critical factor is that the patient status must be monitored or determined constantly. Therefore, realizing a real-time response for service robot may provoke more academic interest not only in the development of hardware but also in human-robot interaction.

5. Conclusion

Although service robots have an infinite amount of latent potential, there is still a possibility that they may fail to meet users' need and cause negative psychological results. The main purpose of the present study was to develop a method of informing a service robot of the users' feelings based on measured CoM-related items. A classifier was designed according to the frequency domain data of CoM during the STS movement. The results showed that the designed classifier had a high probability of discriminating fault classes from others. We also presented a practical and economical experimental method to correlate negative psychological changes to data that a robot can understand.

Acknowledgements

This research was funded by Hirose International Scholarship Foundation (no. J770703161). We gratefully acknowledge the work of past and present members in our laboratory. Lastly, we appreciate all supports and understanding from those people who stood behind us and still are standing by our sides.

Author details

Tianyi Wang[1], Hieyong Jeong[1]* and Yuko Ohno[2]

*Address all correspondence to: h.jeong@sahs.med.osaka-u.ac.jp

1 Department of Robotics & Design for Innovative Healthcare, Graduate School of Medicine, Osaka University, Japan

2 Department of Mathematical Health Science, Graduate School of Medicine, Osaka University, Japan

References

[1] Dall PM, Kerr A. Frequency of the sit to stand task: An observational study of free-living adults. Applied Ergonomics. 2010;**41**(1):58-61. DOI: 10.1016/j.apergo.2009.04.005

[2] Grant PM, Dall PM, Kerr A. Daily and hourly frequency of the sit to stand movement in older adults: A comparison of day hospital, rehabilitation ward and community living groups. Aging Clinical & Experimental Research. 2011;**23**(5-6):437-444. DOI: 10.1007/BF03325239

[3] Guralnik JM, Simonsick EM, Ferrucci L, Glynn RJ, Berkman LF, Blazer DG, Scherr PA, Wallace RB. A short physical performance battery assessing lower extremity function: Association with self-reported disability and prediction of mortality and nursing home admission. Journal of Gerontology. 1994;**49**(2):85. DOI: 10.1093/geronj/49.2.M85

[4] Odding E, Valkenburg HA, Stam HJ, Hofman A. Determinants of locomotor disability in people aged 55 years and over: The Rotterdam Study. European Journal of Epidemiology. 2001;**17**(11):1033-1041. DOI: 10.1023/A:1020006909285

[5] Winograd CH. Lower-extremity function in persons over the age of 70 years as a predictor of subsequent disability. New England Journal of Medicine. 1995;**332**(9):556. DOI: 10.1056/NEJM199503023320902

[6] Geravand M, Korondi PZ, Werner C, Hauer K, Peer A. Human sit-to-stand transfer modeling towards intuitive and biologically-inspired robot assistance. Autonomous Robots. 2017;**41**(3):575-592. DOI: 10.1007/s10514-016-9553-5

[7] Chuy O Jr, Hirata Y, Zhidong W, Kosuge K. Approach in assisting a sit-to-stand movement using robotic walking support system. IEEE International Conference on Intelligent Robots and Systems; Scopus. 2006. DOI: 10.1109/IROS.2006.282007

[8] Burnfield JM, Shu Y, Buster TW, Taylor AP, Mcbride MM, Krause ME. Kinematic and electromyographic analyses of normal and device-assisted sit-to-stand transfers. Gait & Posture. 2012;**36**(3):516-522. DOI: 10.1016/j.gaitpost.2012.05.002

[9] Thomson R, Martin JL, Sharples S. The psychosocial impact of home use medical devices on the lives of older people: a qualitative study. BMC Health Services Research. 2013;**13**(1):467. DOI: 10.1186/1472-6963-13-467

[10] Schenkman M, Berger RA, Riley PO, Mann RW, Hodge WA. Whole-body movements during rising to standing from sitting. Physical Therapy. 1990;**70**(10):638-648. DOI: 10.1016/0094-730X(90)90046-U

[11] Janssen WG, Bussmann HB, Stam HJ. Determinants of the sit-to-stand movement: A review. Physical Therapy. 2002;**82**(9):866-879. DOI: 10.1053/joca.2002.0828

[12] Hesse S, Schauer M, Petersen M, Jahnke M. Sit-to-stand manoeuvre in hemiparetic patients before and after a 4-week rehabilitation programme. Scandinavian Journal of Rehabilitation Medicine. 1998;**30**(2):81. DOI: 10.1055/s-2008-1061832

[13] Schenkman M, Riley PO, Pieper C. Sit to stand from progressively lower seat heights—Alterations in angular velocity. Clinical Biomechanics. 1996;**11**(3):153-158. DOI: 10.1016/0268-0033(95)00060-7

[14] Soangra R, Lockhart TE. A comparative study for performance evaluation of sit-to-stand task with body worn sensor and existing laboratory methods. Biomedical Sciences Instrumentation. 2012;**48**(48):407

[15] Hieyong J, Kenji Y, Michiko K, Shima O, Taishin N, Yuko O. Analysis of difference in center-of-pressure positions between experts and novices during asymmetric lifting. IEEE Journal of Translational Engineering in Health & Medicine. 2016;**4**:1-11. DOI: 10.1109/JTEHM.2016.2599185

[16] Huurnink A, Fransz DP, Kingma I, van Dieën JH. Comparison of a laboratory grade force platform with a Nintendo Wii Balance Board on measurement of postural control in single-leg stance balance tasks. Journal of Biomechanics. 2013;**46**(7):1392-1395. DOI: 10.1016/j.jbiomech.2013.02.018

[17] Clark RA, Bryant AL, Pua Y, Mccrory P, Bennell K, Hunt M. Validity and reliability of the Nintendo Wii Balance Board for assessment of standing balance. Gait & Posture. 2010;**31**(3):307. DOI: 10.1016/j.gaitpost.2009.11.012

[18] Park DS, Lee GC. Validity and reliability of balance assessment software using the Nintendo Wii balance board: Usability and validation. Journal of Neuroengineering & Rehabilitation. 2017;**11**(1):1-8. DOI: 10.1186/1743-0003-11-99

[19] Yamada T, Demura S. The relationship of force output characteristics during a sit-to-stand movement with lower limb muscle mass and knee joint extension in the elderly. Archives of Gerontology & Geriatrics. 2010;**50**(3):46. DOI: 10.1016/j.archger.2009.05.009

[20] Russell JA. A circumplex model of affect. Journal of Personality & Social Psychology. 1980;**39**(6):1161-1178. DOI: 10.1037/h0077714

[21] Swangnetr M, Kaber DB. Emotional state classification in patient-robot interaction using wavelet analysis and statistics-based feature selection. IEEE Transactions on Human-Machine Systems. 2012;**43**(1):63-75. DOI: 10.1109/TSMCA.2012.2210408

[22] Nakamura Y, Sekiguchi A. The chaotic mobile robot. Robotics & Automation IEEE Transactions on. 2001;**17**(6):898-904. DOI: 10.1109/70.976022

[23] Perc M. Nonlinear time series analysis of the human electrocardiogram. European Journal of Physics. 2005;**26**(5):757-768. DOI: 10.1088/0143-0807/26/5/008

[24] Perc M. The dynamics of human gait. European Journal of Physics. 2005;**26**(3):525-534. DOI: 10.1088/0143-0807/26/3/017

[25] Khelifa MA, Boukabou A. Design of an intelligent prediction-based neural network controller for multi-scroll chaotic systems. Applied Intelligence. 2016;**45**(3):1-15. DOI: 10.1007/s10489-016-0793-z

[26] Roldánjiménez C, Bennett P, Cuestavargas AI. Muscular activity and fatigue in lower-limb and trunk muscles during different sit-to-stand tests. PLoS One. 2015;**10**(10):e0141675. DOI: 10.1371/journal.pone.0141675

[27] Yousefi J, Hamiltonwright A. Characterizing EMG data using machine-learning tools. Computers in Biology & Medicine. 2014;**51**(8):1. DOI: 10.1016/j.compbiomed.2014.04.018

[28] Kodba S, Perc M, Marhl M. Detecting chaos from time series. Fractals in the Natural and Applied Sciences. 2005;**33**(33):371-383. DOI: 10.1088/0143-0807/26/1/021

Section 2

Applications

Chapter 4

Robots in Agriculture: State of Art and Practical Experiences

Juan Jesús Roldán, Jaime del Cerro,
David Garzón-Ramos, Pablo Garcia-Aunon,
Mario Garzón, Jorge de León and
Antonio Barrientos

Additional information is available at the end of the chapter

http://dx.doi.org/10.5772/intechopen.69874

Abstract

The presence of robots in agriculture has grown significantly in recent years, overcoming some of the challenges and complications of this field. This chapter aims to collect a complete and recent state of the art about the application of robots in agriculture. The work addresses this topic from two perspectives. On the one hand, it involves the disciplines that lead the automation of agriculture, such as precision agriculture and greenhouse farming, and collects the proposals for automatizing tasks like planting and harvesting, environmental monitoring and crop inspection and treatment. On the other hand, it compiles and analyses the robots that are proposed to accomplish these tasks: e.g. manipulators, ground vehicles and aerial robots. Additionally, the chapter reports with more detail some practical experiences about the application of robot teams to crop inspection and treatment in outdoor agriculture, as well as to environmental monitoring in greenhouse farming.

Keywords: robotics, agriculture, greenhouse, UGV, UAV, multi-robot

1. Introduction

Agriculture can be a field as favourable as industry for the application of automation. The challenges for robots in agriculture are diverse. On the one hand, agricultural environments, in contrast to industrial facilities, are not structured and controlled. On the other hand, industrial processes can be designed by modules to apply specific robots to specific works, whereas the complex tasks of agriculture sometimes cannot be split into simple actions. For the above reasons, agricultural applications require more versatile and robust robots.

In the last years, multiple groups around the world have applied different automation solutions (e.g. sensor networks, manipulators, ground vehicles and aerial robots) to diverse agricultural tasks (e.g. planting and harvesting, environmental monitoring, supply of water and nutrients, and detection and treatment of plagues and diseases). This chapter aims to collect the state of the art about robotics applied to agriculture, as well as to describe more exhaustively some of our practical experiences.

Section 2 addresses the agricultural tasks where the robots can be applied, remarking precision agriculture and greenhouse farming, but covering other works such as automatic planting and harvesting. Section 3 describes the robots applied in agricultural tasks in the state of the art, covering both multi-robot systems, ground and aerial robots. Section 4 summarizes our main experiences in robotics applied to agriculture, which are related to precision agriculture in open fields and environmental monitoring of greenhouses. Finally, Section 5 summarizes the main conclusions acquired from the review of the state of the art and our experience in the research projects.

2. Automation in agriculture

This section reports the state of the art about automation in agriculture. For this purpose, it is organized as follows: Section 2.1 is focused on precision agriculture, Section 2.2 addresses the application of new technologies to greenhouse farming, and Section 2.3 analyzes the proposals for automatic planting and harvesting.

2.1. Precision agriculture

Precision agriculture, also known as precision farming, is a concept of farm management based on the application of different technologies, in order to manage the spatial and temporal variability associated with all aspects of agricultural production. Its main goal is the improvement of both crop performance and environmental quality.

Several authors have confirmed the economic and environmental benefits that are achieved when precision agriculture methodologies are applied. Nonetheless, academic surveys and professional reports show that the rate of adoption of these technologies is still low [1].

Moreover, instead of using precision agriculture as complete concept, most of the deployments reported use these techniques to solve specific needs or to fill important gaps in the knowledge of farmers [2]. Additionally, even though agronomists are playing the leading role in PA development, engineers have worked diligently to provide technologies needed to implement PA practices. Engineering innovations for PA involve development of sensors, controls, and remote-sensing technologies.

Autonomous mobile robots can be used in a variety of field operations. They can be applied to facilitate capturing and processing high quantities of data, and they can provide the capabilities required to operate not only at individual plant level but also at complete field level.

Blackmore and Griepentrog [3] study the autonomous platforms that may be available in the future, which would be used for cultivation and seeding, weeding, scouting, application of fertilizers, irrigation and harvesting.

The most widely used robotic technology in precision agriculture is vehicle guidance and auto-steer systems. The reason is that the economic benefits are easily achievable without requiring the integration of additional components or decision support systems [2]. However, other technologies, especially those related to remote sensing, development of sensors and controls, are also used by teams combining agronomists and engineers. **Table 1** summarizes some of the developments, which use mobile robotics and remote sensing for precision agriculture.

2.2. Greenhouse farming

Greenhouse farming is often a suitable field for applying the technologies of automation, computing and robotics. Some examples of technologies implemented in productive greenhouses are the control of temperature and humidity, the soil preparation and the supply of water and nutrients [10]. The robots can perform some tasks that humans cannot do due to the harsh conditions of greenhouses, such as environmental monitoring and control, crop monitoring, supply and treatment, and pest and disease detection.

The environmental monitoring of greenhouses is interesting not only to control the growth of crops but also to determine the traceability of products. Nowadays, most of the systems used for environmental monitoring of greenhouses are based on wireless sensor networks (WSNs) [11–13]. Nevertheless, the robots are starting to be applied as mobile platforms for sensors [14–16].

Publication	Operation	Technique
[4]	Weeding	Automatic computer vision method for detecting weeds in cereal crops, and differential spraying to control the weeds.
[5]	Field mapping	Creation of 3D terrain maps by combining the information captured with a stereo camera, a location sensor, and an inertial measurement unit, all installed on a mobile equipment platform.
[6]	Field mapping and coverage	An unmanned car-like mobile robot uses a SLAM algorithm to navigate in the agricultural environment while creating a map of such environment.
[7]	Multi-purpose	Design and construction of a multi-purpose mobile ground platform for PA tasks.
[8]	Coverage path planning	Harmony Search (HS) algorithm for finding complex coverage trajectories with a fleet of aerial robots.
[9]	Weed and pest control	A fleet of heterogeneous ground and aerial robots is developed and equipped with innovative sensors, enhanced end-effectors and improved decision control algorithms to cover a large variety of agricultural situations.

Table 1. Precision agriculture: main operations and techniques.

Greenhouses can be considered complex multiple-input multiple-output systems [17]. The literature collects multiple proposals for modelling and controlling the conditions of greenhouses [18]. Some of them obtain the models of greenhouses applying analytical equations (e.g. mass and energy balances) [19], whereas the rest identifying process models (e.g. neural networks or fuzzy sets) [20]. A review of these models determined the input, output, and disturbance variables described below.

The input variables allow to actuate the greenhouses and change the environmental conditions. The most relevant variables considered by literature are the ventilation [21], heating [22], fogging [23], shading and CO_2 injection [24] systems. The ventilation systems control the exchange of air between greenhouse and environment, which has an impact on the air temperature, humidity and composition. The heating systems are used to compensate the heat losses and keep the temperatures in the adequate range. The fogging systems spray water into the air to increase the humidity and reduce the temperature. The shading systems control the irradiation of the covers to avoid the overheating of the greenhouse. Finally, the CO_2 injection systems are used to promote the photosynthesis of the plants.

The output variables define the state of greenhouse can be measured by the appropriate sensors and are the target of climate control. The most relevant variables collected by literature are the air temperature, air humidity, solar radiation and CO_2 concentration. In addition, there are some variables that have influence on the state of greenhouse and should be measured and controlled. These disturbances are the external temperature, external humidity, wind speed, wind direction, external CO_2 concentration, cover temperature, crop temperature and ground temperature.

Table 2 collects the relevant variables for the environmental monitoring of greenhouses [25], as well as the appropriate sensors to measure them and their possible application in robots.

Another task of greenhouse farming where the robots can perform an important role is the crop inspection and treatment. The detection of weeds [26], pests [27] and diseases [28] is possible through direct and indirect methods. The direct methods are based on acquiring RGB [29] and 3D [30] images and applying computer vision techniques. The indirect ones require to take samples in the greenhouse and to analyze them in the laboratory. The ground robots can be used to apply treatments and fertilizers to the crops [31–33], in order to improve the precision and rationalize the products.

Planting and harvesting are seasonal tasks that require a considerable amount of work. The literature contains some proposals to automatize these tasks [34–36]. These proposals consider different types of robots (ground mobile and rail robots), sensors (mainly, RGB and 3D cameras and laser scanners) and effectors (manipulators and graspers).

2.3. Seeding and harvesting

Regarding the point of view of service robotics, Onwude et al. [37] attempt to evaluate the application of agricultural mechanization and its present technologies and limitations for the large-scale purposes. The study shows an increasing level of technological advancement in field and crop mapping, soil sampling, mechanical seeders and harvesters in agricultural robots. Sharing the same approach, Kester et al. [38] show the future trends and the likely

Variable	Sensor	Application
Radiation (absorbed)	Net radiometer	AR, GR, FS
Radiation (solar)	Pyranometer	AR, GR, FS
Air temperature	Resistance temperature device	AR, GR, FS
	Thermocouple	AR, GR, FS
	Thermistor	AR, GR, FS
Surface temperature	Infrared	AR, GR, FS
	Thermocouple	GR, FS
Substrate temperature	Resistance temperature device	GR, FS
	Thermocouple	GR, FS
	Thermistor	GR, FS
Air humidity	Capacitance hygrometer	AR, GR, FS
	Condensation hygrometer	AR, GR, FS
	Psychrometer	AR, GR, FS
Ground humidity	Electrical conductivity meter	GR, FS
Carbon dioxide in air	Non-dispersive infrared sensor	AR, GR, FS
pH	pH probe	GR, FS

AR, air robot; GR, ground robot; FS, fixed sensor.

Table 2. Environmental monitoring of greenhouses: variables, sensors and robots based on Ref. [16].

adoption of automated farming machinery. The results of this study point out a growing interest in autonomous and semi-autonomous systems for reducing the highest workload operations: tillage, seeding and harvesting.

In order to support the growing development of seeding and harvesting robots, new strategies are proposed for driving autonomous mobile robot in the associated scenarios. Ko et al. [33] makes a small review about common techniques for robot navigation in greenhouses and proposes a methodology based on machine learning. Due to the importance of the manipulation tasks in seeding, transplanting and harvesting processes, the manipulator motion strategy is also discussed by several authors. However, they usually do not specify path planning algorithms; the most common approach is the direct displacement towards the end-effector desired position by position-based control [39] and visual feedback control [40]. The task planning strategies are only studied by a few researchers. Commonly, the harvesting task is limited to pick one fruit, while the planning required for picking the rest is avoided. Nevertheless, the problem can be studied from two perspectives: the coverage path planning [41] for picking all the fruits in a scene or the time minimization [42] for moving from a fruit to another. The obstacle detection and avoidance are studied by an equally low number of authors. A great complexity is added to the solution due to the obstacle recognition in addition to the path planning algorithms. A few approaches are based in the obstacle detection with collision sensors in the end effector [43], the obstacle recognition by Light Imaging, Detection and Ranging (LIDAR) [44] and vision techniques [45].

A review about the use of technologies for automated activities in greenhouses is presented in Ref. [46], showing an increasing implementation of wireless systems for environmental measurements. Additionally, the study shows that a considerable number of research works are aimed to develop robotic systems for fruit picking and extraction. Furthermore, the research community has put much effort on developing techniques for robust fruit recognition; moreover, there is a high necessity for improving the picking capabilities of transplanting and harvesting robots in order to move towards a commercial application. A review on vision control techniques and their potential applications in fruit or vegetable harvesting robots is presented in Ref. [47]. The fruit identification and localization are the most common problem studied by the authors. Like the fruit ripeness identification [48, 49], a great number of approaches are based on RGB cameras [50], as well as colour and shape recognition [51]. The multi-spectral lightning is less studied despite more information can be acquired with this kind of technology [52]. The next level of complexity involves the implementation of stereo vision systems and LIDAR for calculating the fruit position in a 3D space [53].

This objective can only be accomplished by the diversification and specialization of the robotic systems. With the aim of getting better results in harvesting tasks, newer and more precise sensors are needed. In Ref. [36], the authors review the modern sensor systems used in semi or fully automated robotic harvesting. Their research shows how the integration of several kinds of technologies and sensor fusion can improve the precision in fruit recognition and localization activities.

Some interesting publications to be considered for a further review of seeding, transplanting and harvesting robotics can be found in Refs. [34, 54].

3. Agricultural robots

This section reviews the different types of robots and payloads that are applied to agricultural tasks. For this purpose, it is organized as follows: Section 3.1 analyzes the aerial robots, Section 3.2 the ground robots, Section 3.3 some special robots that are not conventional ground or aerial vehicles and Section 3.4 the multi-robot systems.

3.1. Aerial robots

The Association for Unmanned Vehicle Systems International (AUVSI) published an economical report in 2013 [55], which emphasized the future impact of the civil use of the unmanned aerial vehicles (UAVs) on the USA economy. This document highlights two markets over multiple areas: public safety and precision agriculture. It concludes that by far and above, this last market will be the largest in the next decade in terms of economy and jobs.

In the last decades, collecting data from agricultural holdings has been carried out mainly by manned aerial vehicles, satellites or directly specialized experts at floor level [56]. The availability of these methods has some limitations, such as the presence of clouds, the long data delivery times, the need for special permissions and the prizes of some products. In contrast, the UAVs can be deployed efficiently, can carry multiple types of sensors, do not require very restrictive permissions and are becoming a cost-effective alternative. These advantages have been reinforced with the rise of the vertical take-off and landing (VTOL) vehicles and more specifically the quadcopters (a good example can be seen in Ref. [57]).

One of the first applications of UAVs in precision agriculture was their use to measure the water stress in agricultural holdings. Nowadays, the UAVs are equipped with thermal and hyperspectral cameras, as well as fluorescence sensors [58]. An interesting experiment is reported in Ref. [59], where the authors produced a controlled deficit of irrigation to generate a gradient of water stress in a citrus orchard. They compared the data obtained by the micro-thermal and hyperspectral cameras boarded on a fixed-wing unmanned aircraft with the measurements on the leaves, validating the aerial methods to measure the water stress. Similar works can be found in Refs. [60–62], which show the feasibility and benefits of the aerial thermal imagery to improve the irrigation.

Another use of aerial vehicles in precision agriculture is the monitoring of crops to predict yields, properly calculate the amount of fungicides and fertilizers and detect pathogen. In Ref. [63], a RGB camera was used to estimate the biomass of a barley exploitation under two different nitrogen treatments. The results were cross-validated with five different crop surface models based on the height of the plant. Additionally, in Ref. [64], techniques of clustering were applied to estimate the biomass of a wheat exploitation through the RGB images. The method was enhanced by measuring the height of the plant directly on ground. These studies demonstrate the feasibility, applicability and precision of this tool.

Another interesting parameter for monitoring is the leaf area index (LAI), defined as the ratio of leaf surface to unit ground area. In Ref. [65], the authors estimate the LAI of maize, potato and sunflower fields by using a hyperspectral camera and inverting the solar model of the canopies. In [66], they used a visual 3D modelling technique to estimate the LAI in a vineyard, obtaining around 57% less precision. Although this method seems to be nowadays less accurate, it is quick, inexpensive and practical compared to other methods. In Ref. [67], they use four cameras and overlap the corresponding images to measure the LAI over wheat and rapeseed crops. They achieve a very good correlation in comparison to the measurements at floor level, despite that they highlight the dependency of the method on the light conditions.

Additionally, the UAVs are a practical tool for the prevention of diseases in agricultural holdings. The work reported in Ref. [68] uses a UAV and a manned aircraft equipped with multispectral cameras. It applies four classification algorithms to detect the prints of the diseases and compare the results with the ground truth. Similarly, the work described in Ref. [69] proposes a new airborne visual sensor to detect diseases on the leaves on citrus trees, and the experiments show an accuracy of 93%. These methods have been used not only for citrus trees but also for olives [70], avocado trees [71], potatoes [72] and grapevines [73], among others.

Although this use is not extended around the world due to the restrictions in many countries, there have been some trials and investigations with UAVs spraying locally fertilizers and pesticides [74]. For instance, in Ref. [75], a net of pesticide sensors are deployed over a virtual field, whereas a UAV sprays the chemicals where they are needed. The sensors provide a feedback to the UAV, which adjusts the optimal route and minimizes the waste of pesticides. A more practical and exhaustive study was carried out in Ref. [76], where different spraying parameters are tested in a field with rice crops, such as operation height and velocity, and the deposited volume at different heights of the plants.

As we have seen, the UAVs equipped with hyperspectral and thermal cameras are an affordable and practical system to improve the performance of agricultural holdings, reducing costs and increasing productivity. With the development of new sensors and the establishment of new legal frameworks, their expansion will continue in the next decades.

3.2. Ground robots

The systematic design methods assist researchers in design choices, whereas the economic analysis considers allowable cost of a system. Only a few authors report design processes based in requirement engineering. The decisions about the hardware have influence on the robot performance, the complexity of algorithms and, eventually, the system costs. Some approaches for ground robots are reviewed with special interest in the platform, manipulator and end-effector implemented in several agricultural applications.

Regarding to the platform, custom mobile robots are the most common choice [77]. The robustness of crawler and caterpillar platforms [78, 79] is described by several authors, but the wheeled robots [80] are also shown as good choices due to their simplicity. All these platforms can integrate approaches such as GPS [43], odometry [50], line guidance [81], path plans [82] and manned approaches [83] as navigation strategy. Other robots take advantage of irrigation pipes or rails in the field [84, 85] to achieve their displacement.

The most common configuration of the manipulators is near 3 DoF (Degrees of Freedom) in a custom-made development [86, 87]. This situation can be associated to the high cost of implementations involving commercial manipulators [39]. Although the manipulators are commonly custom made, neither analysis nor explanation for the number of DOF selected is usually detailed by the authors. The manipulator movements are just simplified, and the high numbers of DOF are avoided.

Finally, most end-effectors are designed to operate with two fingers [44, 79], since most of the grasps can be performed by them, and they are the smallest suitable mechanical architectures for grasping hand devices. In addition, grasping is commonly achieved by suction grippers [88]; for that reason, the end effectors are mainly actuated by electrical and pneumatic systems. A great number of the end effectors are custom made [89, 90]; this design preference can be associated to the huge diversity of tasks to be performed, as well as the several kinds of fruits and vegetables to be handled. Additionally, it is a common practice trying to solve the problems originated by the mobile platform and the manipulator through the end-effector behaviour; this situation leads to greater customization of the grippers.

For comprehensive reviews of the agricultural robotics literature and further exemplification of their ground robot applications, we refer the reader to the readings [91, 92].

3.3. Special robots

The traditional ground robots present several limitations related to the constraints of wheeled and caterpillar motion. Additionally, the use of aerial robots is not always possible, especially when the task should be performed on surface or indoor. This section reports some cases in the literature where robots with alternative locomotion systems are applied to agricultural tasks.

Sphere robots are systems whose movements are produced by instability. This type of robots is used in several applications and scenarios, such as exploration [93] and surveillance [94]. ROSPHERE is described by their authors as a new low-cost spherical robot for measuring soil temperature and moisture in precision agriculture [95]. In comparison to wheeled robots with similar size and capabilities, the ROSPHERE is much lighter and robust in irregular terrains.

Several robots are designed by bioinspiration, which tries to replicate the biological evolution. For instance, the engineers have noticed that hexapod insects are able to walk in all terrains, so they are replicating their physiognomies and walking patterns. This is the case of Prospero, which is a prototype of hexapod robot that can plant, tend and harvest autonomously [96]. Another example are the RoboBees of Harvard University, which are micro-aerial robots inspired in bees [97]. These robots are being applied for distributed environmental monitoring and assistance to crop pollination.

3.4. Multi-robot systems

Sometimes, single robots are not able to perform some complex tasks (e.g. those that require coordinated actions in multiple locations) or to perform simple tasks in the required time (e.g, when the tasks must be performed in large areas). In these cases, robot teams might provide some advantages over single robots, such as their effectiveness, efficiency, flexibility and fault tolerance.

Most of the cases of robot teams for agricultural tasks reported by the literature are homogeneous. Using a fleet of UAVs instead of a single UAV for collecting data in large areas is common, and there are multiple techniques for area distribution and path planning [98]. For instance, a team of small UAVs with low-cost cameras can be applied to control the exploitation and management of water [99], obtaining the same results than a single UAV equipped with a better camera and providing the operation with more robustness.

Nevertheless, some agricultural applications require heterogeneous robot teams. The most common situation is when the task consists of operations that are more effective from the air and others that are more effective from the ground. For instance, aerial robots are able to efficiently cover large fields taking pictures and collecting data (e.g. distribution of water or location of weeds), whereas ground robots can actuate on the crops with more robustness and precision (e.g, watering or applying treatments) [9].

Therefore, the heterogeneous multi-robot systems often combine the advantages and compensate the drawbacks of different robots. The Section 4 describes with more details two different multi-robot systems applied to two different agricultural scenarios: outdoor agriculture and greenhouse farming.

4. Practical experience

This section reports some of the experiences of the Robotics and Cybernetics Group (RobCib) in the context of robotics applied to agriculture. It is organized as follows: Section 4.1 summarizes the participation in the Robot Fleets for Highly Effective Agricultural and Forestry Management (RHEA) Project, whereas Section 4.2 describes the use of multiple robots for the environmental monitoring of greenhouses.

4.1. RHEA project

A good example of precision agriculture is the RHEA (Robot Fleets for Highly Effective Agricultural and Forestry Management) Project. It was carried out under the seventh framework program of the European Commission and identified as NMP-CP-IP 245986-2. RHEA activities finished on 31 July 2014. The project was focused on designing, developing and testing of a new generation of robotic systems for both chemical and physical (mechanical and thermal) effective weed management in the context of agriculture and forestry.

The use of pesticides in agriculture helps to improve yields and to prevent crop losses. Nevertheless, pesticides include active ingredients that have adverse impacts on the environment and habitats. According to the "Agriculture, forestry and fishery statistics" [100] online publication of Eurostat, the sale of pesticides in European Union member states amounted to close to 400,000 tonnes in 2014. Due to their potential toxicity, the application of pesticides is strictly controlled by EU legislation since 1991 and previously by national regulations. Although the return of pesticides (the crops saved from pests and diseases) is approximately four times the investment [101], the indirect costs (the impact on human health and environment) are estimated to total approximately $10,000 million per year in the United States [101].

The farmers usually apply these pesticides by using traditional sprayers that distribute them uniformly over the complete field. Therefore, the aim of RHEA Project was to support the farmers by reducing the amount of pesticides applied without reducing the effectiveness of the treatments. This objective was reached by applying the pesticides with high precision only where it is required. This solution not only prevents the pernicious effects of the pesticides but also reduces drastically the economic cost of treatment.

RHEA scope covers a large variety of European products, such as agriculture wide row crops (processing tomato, maize among others), close row crops (winter wheat and winter barley) and forestry woody perennials (walnut trees, almond trees, olive groves and multi-purpose open woodland).

The project is based on the cooperation among aerial and ground vehicles to perform precision agriculture tasks, namely weeds removal and trees fumigation. A complete description of the system as well as the result of the project can be found at Ref. [9].

Ground units are based on small *Case New Holland Industrial* tractors with some modifications.

http://dx.doi.org/10.5772/intechopen.69874

In order to provide the system with the required autonomy, several kinds of sensors and actuators were deployed into the systems.

A high precision Global Navigation Satellite System (GNSS) for autonomous outdoor navigation was used to allow the control system to accurately steer the robots to work on wide-row crops (with 0.75 m-spaced rows). Additionally, a ground perception system was used to discriminate weeds from crops while travelling along crop rows, as well as a real-time tree canopy detection system.

The tractors were also endowed with three kinds of actuators, namely a patch sprayer aiming at reducing herbicide use by approximately 75%, a canopy sprayer to reduce the use of pesticide in canopy spraying by approximately 50% and a mechanical/thermal tool to destroy 90% of the detected weeds.

Additionally, the tractors were provided with a communication equipment, a sustainable energy system and a safety systems for humans and animals detection. **Figure 1** shows the ground units with the mentioned actuators.

In order to plan the activities of the ground units, the RHEA concept includes the support of aerial imagery by using drones. Thus, the system used a fleet of last-generation hexacopters provided by *Air Robot* that rely on high payload and extraordinary stability (shown in **Figure 2**). These features allow taking steady pictures with high-quality cameras in large open fields. As result, high-resolution images of the open field are obtained, in order to provide the ground units with the locations of weeds to remove them.

Usually, weeds are not uniformly spread over the fields but are located in patches. For this reason, the first step of the mission is to obtain high-precision aerial images to locate the patches in the field. The location of the patches will allow defining optimal paths for the tractors. This task requires a thorough planning of the flights, which was our main task in this project. Several area coverage planning techniques were applied taking into account spatial and temporal requirements, in order to generate optimal and safe flight plans for the drones. A complete description of the aerial mission can be found in Refs. [102, 103].

Figure 1. Ground units of RHEA project.

Figure 2. Aerial units of RHEA project.

The drones were equipped with two high-resolution cameras (4704 × 3136) mounted on a gimbal system. A multi-spectral device that includes visible and near infrared (NIR) channels was selected to maximize the robustness under different light conditions. Moreover, in order to preserve the complete colour information, the solution uses a coupling of two commercial still cameras: one of them modified to provide the NIR channel. The flights were performed with an elevation of 60 m to obtain a resolution of 1 cm per pixel with each image covering approximately 40 × 30 m and overlapping 60% with the consecutive ones.

RHEA provides the farmer with a ground station with a graphical user interface (GUI). This interface allows the farmer to create and launch the mission. Thus, the operator has to define the area by entering the points that limit the field.

Once the limits of the field have been established, the system creates a flight plan for each robot of the fleet. After testing several optimization techniques, back and forth motion movements are applied for planning the fleet trajectories, in order to increase the situational awareness of the operator. Although the drones are able to autonomously perform take-off and landing operations, the pilot needs to provide altitude commands to ensure a safe operation.

Later, a mosaicking procedure is developed: colour and NIR images from the two cameras are joined in a unique four-channel picture. An approach based on the Fourier-Mellin (FM) transform was successfully developed and tested. This approach identifies rotation, translation and scale changes between images by means of Fourier spectrum analysis. To cope with large-sized camera images, which imply non-linear transformations, the original images are partitioned into a set of small image portions, where the FM identification process is executed iteratively. Then, a global homographic transformation model is computed including lens radial distortion. A registration accuracy of 0.3 pixels is obtained [17]. After obtaining a global image of the fields, the patch detection should be performed.

Detecting weeds when crop and weed plants are at early phenological stages is a challenge. The proposal of RHEA overcomes it by using high-spatial-resolution imagery and object-based image analysis (OBIA), taking into account the relative position of the weeds to the crop lines, so that every plant that is not located on the crop row is considered a weed [104]. As a result, a weed patch map is created using a grid of 0.5 m. The size of this grid is customizable according to the requirements of the herbicide-spraying machinery.

Later, the Ground Mission Planner is executed by the operator in the base station. It determines the configuration of the ground vehicles (number and type of vehicles), as well as the plan for each one to efficiently apply the treatment. Simulated annealing and basic genetic algorithms are used to find the optimal solution that minimizes either the task cost or the time required cost, whereas a non-dominated sorting genetic algorithm (NSGA-II) is employed as a proper approach for simultaneously minimizing both criteria [105].

Once defined the optimal paths, the ground mission starts. During the mission, the ground perception system detects weeds, both inter and intra row. This system is based on a SVS-VISTEK camera connected to the high-level decision-making system computers for acquiring images and running the relevant vision algorithms [106]. The operation speed of Unmanned Ground Vehicles (UGVs) was fixed at 0.83 m/s, and the region of interest (ROI) for the ground perception system was defined to be 3-m wide and 2-m long located in front of the UGV. Weed detection relies on the spatial identification of crop rows. Thus, determination of crop-row positions with respect to the UGV becomes a key task for weed patch detection and UGV guiding. Weed detection system also generates the orders to the sprayers to activate precisely when nozzle passes over the weeds.

4.2. Environmental monitoring in greenhouses

As mentioned above, one of the agricultural applications where the robots can work is the environmental monitoring of greenhouses. This task cannot be performed by humans, because it requires a continuous work under the harsh conditions of greenhouses. The alternatives of the robots in this task are the fixed sensors, which are not able to capture the spatial variability of the environmental conditions, and the sensor networks, which cannot be moved during the operation to the points of interest.

The proposed solution is a multi-robot system that measures the environmental variables and collects their spatial and temporal variability. This information is the key to control the conditions of the crops, which determine both the productivity and the quality of greenhouse. As shown in **Figure 3**, the multi-robot system is split into small teams with ground and aerial units that work in specific areas. A base station controls the mission, coordinating the actions of the robots, as well as collecting and storing their measurements.

The first work used a mini-UAV to measure air temperature, humidity, luminosity and carbon dioxide concentration [5]. The sensors shown in **Table 3** were selected to measure these variables regarding their size, weight, range, resolution and cost. These sensors were integrated by means of a Raspberry Pi computer, which collects the measurements and sends them to the base station via Wi-Fi network.

The main contributions of this work were the aerodynamics study of quadcopter and the real experiments with the sensory system. The results validated the use of quadcopter as sensory system and determined the optimal location of sensors: the centre and top of the quadrotor frame. A subsequent work developed a chamber to measure the concentration of gases minimizing the influence of propellers and obtaining less errors and fluctuations [107].

The second work introduced a medium-size UGV to measure ground temperature and humidity [6]. Specifically, we used the distance infrared temperature sensor and the contact conductivity humidity sensor collected by **Table 3**. Additionally, the work studied the path planning and following strategies to cover the greenhouse in the minimum time.

The aerial and ground robots have different strengths and weaknesses. The aerial robots can move fast and agile through the corridors and reach any point in the 3D space. On the other hand, the ground robots have more autonomy to cover the greenhouse and more robustness to avoid accidents. Therefore, a heterogeneous team can take advantage of the potential of both types of robots [7].

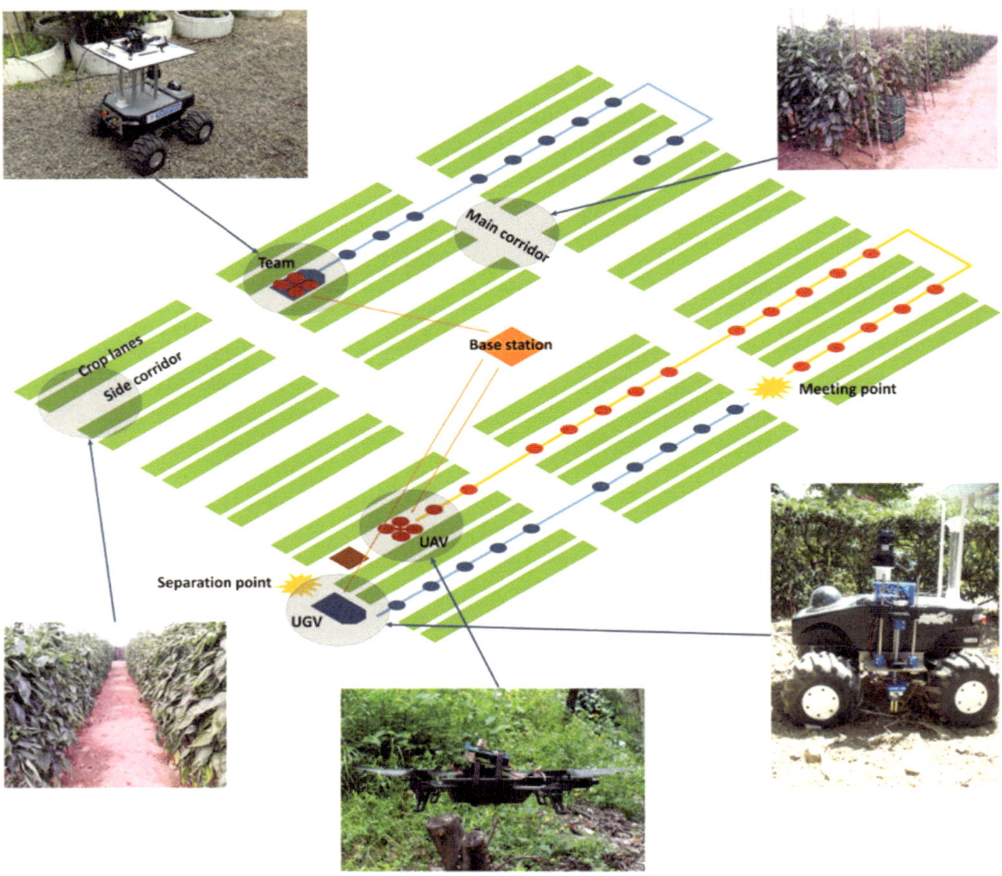

Figure 3. Scheme of multi-robot system for environmental monitoring of greenhouses.

Variable	Sensor	Robot	Controller
Air temperature	RHT03	UAV/UGV	Raspberry Pi
Ground temperature	MLX90614	UGV	Arduino
Air humidity	RHT03	UAV/UGV	Raspberry Pi
Ground humidity	SEN92355P	UGV	Arduino
Luminosity	TSL2561	UAV/UGV	Raspberry Pi
CO_2 concentration	MG811	UAV/UGV	Raspberry Pi

Table 3. Experience in environmental monitoring in greenhouses.

The team strategy of the multi-robot system is the following: the UGV carries the UAV on a platform while it develops its tasks, and when it is required, the UAV takes-off, performs some tasks and lands on the UGV. In this manner, the multi-robot system can avoid the obstacles in the corridors, as well as find the source of anomalous measurements.

The main future challenges of this research line are related to the navigation of the UAVs and the autonomy of UGV and UAV. The navigation of UAVs in the greenhouse is a challenge because the scenario is closed and has high occupancy. The autonomy of both robots is required for the continuous operation and needs better batteries and charging systems.

5. Conclusions

In recent years, the robots have found their own place in agriculture. This chapter addresses the main fields of application (precision agriculture, greenhouse farming, and seeding and harvesting), analyzes the aerial, ground and special robots used in these applications, and describes two research projects related to precision agriculture and greenhouse farming. The main conclusions of these sections are summarized below.

Precision agriculture seeks to apply multiple technologies to acquire knowledge about the spatial and temporal variability of crops. Among other technologies, the use of aerial robots to build maps of the fields and detect weeds or irrigation deficits, and the application of ground robots to apply accurate treatments to plants must be remarked. In addition, greenhouse agriculture has included robots for multiple tasks, such as the monitoring of environmental variables, which is important for the control of the conditions of crops, and the watering and spraying of plants. Finally, although the use of robots for seeding and harvesting is in an earlier step of development, a series of techniques of perception, positioning and grasping have been developed.

The most widely used robots in agriculture are UAVs and UGVs. Aerial robots are usually applied to acquire information of the fields by taking advantage of the altitude. Although the first agricultural aerial robots were fixed-wing UAVs, nowadays the multi-rotors are more popular due to their flexibility. On the other hand, ground robots are usually used to act on the crops. The most common configurations are wheeled and caterpillar robots, and some of the most relevant issues are the selected location and navigation algorithms. Nevertheless, other designs such as spherical or bio-inspired robots are gaining interest, as well as the use of multi-robot systems that can go further than single-robot ones.

Finally, the chapter summarizes two different projects that address the application of robots in agriculture. The RHEA project uses aerial robots to locate weeds within the fields and ground robots to apply localized treatments on them. The other project introduces ground and aerial robots in greenhouses to take measurements of environmental variables. A series of lessons have been learned from these experiences, such as the potential of robots as moving sensory and actuation systems, the difficulties of navigation in unstructured scenarios, the power of cooperation with heterogeneous fleets and the limitations imposed by the autonomy of robots.

Acknowledgements

The research leading to these results has received funding from the RoboCity2030-III-CM project (Robótica aplicada a la mejora de la calidad de vida de los ciudadanos. Fase III; S2013/MIT-2748), funded by Programas de Actividades I+D en la Comunidad de Madrid and the Structural Funds of the EU, from the DPI2014-56985-R project (Protección robotizada de infraestructuras críticas) funded by the Ministerio de Economía y Competitividad of Gobierno de España, and from the NMPCP-IP 245986-2 RHEA project (Robot Fleets for Highly Effective Agriculture and Forestry Management) sponsored by the European Commission's Seventh Framework Programme.

Author details

Juan Jesús Roldán, Jaime del Cerro, David Garzón-Ramos, Pablo Garcia-Aunon, Mario Garzón, Jorge de León and Antonio Barrientos*

*Address all correspondence to: antonio.barrientos@upm.es

Robotics and Cybernetics Research Group, Centre for Automation and Robotics (CAR: UPM-CSIC), Madrid, Spain

References

[1] Pierpaoli E, Carli G, Pignatti E, Canavari M. Drivers of precision agriculture technologies adoption: A literature review. Procedia Technology. 2013;**8**:61-69

[2] McBratney A, Whelan B, Ancev T, Bouma J. Future directions of precision agriculture. Precision Agriculture. 2005;**6**(1):7-23

[3] Blackmore S, Griepentrog HW. A future view of precision farming. In: Berger D, et al., editor. Proceedings of the PreAgro Precision Agriculture Conference. Muncheberg, Germany: Center for Agricultural Landscape and Land Use Research (ZALF); 2002. pp. 131-145

[4] Tellaeche A, BurgosArtizzu XP, Pajares G, Ribeiro A, Fernández-Quintanilla C. A new vision-based approach to differential spraying in precision agriculture. Computers and Electronics in Agriculture. 2008;**60**(2):144-155

[5] Rovira-Más F, Zhang Q, Reid JF. Stereo vision three-dimensional terrain maps for precision agriculture. Computers and Electronics in Agriculture. 2008;**60**(2):133-143

[6] Cheein FA, Steiner G, Paina GP, Carelli R. Optimized EIF-SLAM algorithm for precision agriculture mapping based on stems detection. Computers and Electronics in Agriculture. 2011;**78**(2):195-207

[7] Nielsen SH, Jensen K, Bøgild A, Jørgensen OJ, Jacobsen NJ, Jæger-Hansen CL, Jørgensen RN. A low cost, modular robotics tool carrier for precision agriculture research. In 11th International Conference on Precision Agriculture; 15-18 July 2012; Indianapolis, United States. International Society of Precision Agriculture; 2012

[8] Valente J, Del Cerro J, Barrientos A, Sanz D. Aerial coverage optimization in precision agriculture management: A musical harmony inspired approach. Computers and Electronics in Agriculture. 2013;**99**:153-159

[9] Gonzalez-de-Santos P, Ribeiro A, Fernandez-Quintanilla C, Lopez-Granados F, Brandstoetter M, Tomic S, et al. Fleets of robots for environmentally-safe pest control in agriculture. Precision Agriculture. 2016;1-41

[10] Martínez M, Blasco X, Herrero JM, Ramos C, Sanchis J. Monitorización y control de procesos. una visión teórico-práctica aplicada a invernaderos. RIAII. 2005;**2**(4):5-24

[11] Pawlowski A, Guzman JL, Rodríguez F, Berenguel M, Sánchez J, Dormido S. Simulation of greenhouse climate monitoring and control with wireless sensor network and event-based control. Sensors. 2009;**9**(1):232-252

[12] Pahuja R, Verma HK, Uddin M. A wireless sensor network for greenhouse climate control. IEEE Pervasive Computing. 2013;**12**(2):49-58

[13] Cama-Pinto A, Gil-Montoya F, Gómez-López J, García-Cruz A, Manzano-Agugliaro F. Wireless surveillance system for greenhouse crops. Dyna. 2014;**81**(184):164-170

[14] Roldán JJ, Joossen G, Sanz D, del Cerro J, Barrientos A. Mini-UAV based sensory system for measuring environmental variables in greenhouses. Sensors. 2015;**15**(2): 3334-3350

[15] Ruiz-Larrea A, Roldán JJ, Garzón M, del Cerro J, Barrientos A. A UGV approach to measure the ground properties of greenhouses. In: Robot 2015: Second Iberian Robotics Conference. Springer International Publishing; 2016. pp. 3-13

[16] Roldán JJ, Garcia-Aunon P, Garzón M, de León J, del Cerro J, Barrientos A. Heterogeneous Multi-Robot system for mapping environmental variables of greenhouses. Sensors. 2016;**16**(7):1018

[17] Zeng S, Hu H, Xu L, Li G. Nonlinear adaptive PID control for greenhouse environment based on RBF network. Sensors. 2012;**12**(5):5328-5348

[18] Van Henten EJ. Greenhouse climate management: An optimal control approach (No. 631.34 H4). 1994

[19] Fourati F, Chtourou M. A greenhouse control with feed-forward and recurrent neural networks. Simulation Modelling Practice and Theory. 2007;**15**(8):1016-1028

[20] Rodríguez F, Berenguel M, Guzmán JL, Ramírez-Arias A. Modeling and Control of Greenhouse Crop Growth. London, UK: Springer; 2015

[21] Sethi VP, Sharma SK. Survey of cooling technologies for worldwide agricultural greenhouse applications. Solar Energy. 2007;**81**(12):1447-1459

[22] Sethi VP, Sharma SK. Survey and evaluation of heating technologies for worldwide agricultural greenhouse applications. Solar Energy. 2008;**82**(9):832-859

[23] Arbel A, Barak M, Shklyar A. Combination of forced ventilation and fogging systems for cooling greenhouses. Biosystems Engineering. 2003;**84**(1):45-55

[24] Zhang Z, Liu L, Zhang M, Zhang Y, Wang Q. Effect of carbon dioxide enrichment on health-promoting compounds and organoleptic properties of tomato fruits grown in greenhouse. Food Chemistry. 2014;**153**:157-163

[25] Both AJ, Benjamin L, Franklin J, Holroyd G, Incoll LD, Lefsrud MG, Pitkin G. Guidelines for measuring and reporting environmental parameters for experiments in greenhouses. Plant Methods. 2015;**11**(1):1

[26] Slaughter DC, Giles DK, Downey D. Autonomous robotic weed control systems: A review. Computers and Electronics in Agriculture. 2008;**61**(1):63-78

[27] Chung BK, Xia C, Song YH, Lee JM, Li Y, Kim H, Chon TS. Sampling of Bemisia tabaci adults using a pre-programmed autonomous pest control robot. Journal of Asia-Pacific Entomology. 2014;**17**(4):737-743

[28] Fang Y, Ramasamy RP. Current and prospective methods for plant disease detection. Biosensors. 2015;**5**(3):537-561

[29] Li Y, Xia C, Lee J. Detection of small-sized insect pest in greenhouses based on multifractal analysis. Optik-International Journal for Light and Electron Optics. 2015;**126**(19):2138-2143

[30] Xia C, Wang L, Chung BK, Lee JM. In situ 3d segmentation of individual plant leaves using a rgb-d camera for agricultural automation. Sensors. 2015;**15**(8):20463-20479

[31] Sammons PJ, Furukawa T, Bulgin A. Autonomous pesticide spraying robot for use in a greenhouse. In: Proceedings of the Australian Conference on Robotics and Automation; 5 December 2005; Sydney, Australia. 2005. pp. 1-9

[32] Belforte G, Deboli R, Gay P, Piccarolo P, Aimonino DR. Robot design and testing for greenhouse applications. Biosystems Engineering. 2006;**95**:309-321

[33] Ko MH, Ryuh BS, Kim KC, Suprem A, Mahalik NP. Autonomous greenhouse mobile robot driving strategies from system integration perspective: Review and application. IEEE/ASME Transactions on Mechatronics. 2015;**20**:1705-1716

[34] Bac CW, Henten EJ, Hemming J, Edan Y. Harvesting robots for High-value crops: State-of-the-art review and challenges ahead. Journal of Field Robotics. 2014;**31**(6):888-911

[35] Bachche S. Deliberation on design strategies of automatic harvesting systems: A survey. Robotics. 2015;**4**(2):194-222

[36] Zujevs A, Osadcuks V, Ahrendt P. Trends in robotic sensor technologies for fruit harvesting: 2010-2015. Procedia Computer Science. 2015;**77**:227-233

[37] Onwude DI, Abdulstter R, Gomes C, Hashim N. Mechanisation of large-scale agricultural fields in developing countries—A review. Journal of the Science of Food and Agriculture, 2016, vol. 96, no 12, p. 3969-3976

[38] Kester C, Griepentrog HW, Hörner R, Tuncer Z. A survey of future farm automation—A descriptive analysis of survey responses. In: Precision Agriculture'13. Wageningen Academic Publisher, Wageningen, Netherlands, 2013

[39] Hayashi S, Shigematsu K, Yamamoto S, Kobayashi K, Kohno Y, Kamata J, Kurita M. Evaluation of a strawberry-harvesting robot in a field test. Biosystems Engineering. 2010;**105**(2):160-171

[40] Foglia MM, Reina G. Agricultural robot for radicchio harvesting. Journal of Field Robotics. 2006;**23**(6-7):363-377

[41] Baeten J, Donné K, Boedrij S, Beckers W, Claesen E. Autonomous fruit picking machine: A robotic apple harvester. In: Field and Service Robotics. Berlin Heidelberg: Springer, Berlin, Germany, 2008, pp. 531-539

[42] Reed JN, Miles SJ, Butler J, Baldwin M, Noble R. AE—Automation and emerging technologies: Automatic mushroom harvester development. Journal of Agricultural Engineering Research. 2001;**78**(1):15-23

[43] De-An Z, Jidong L, Wei J, Ying Z, Yu C. Design and control of an apple harvesting robot. Biosystems Engineering. 2011;**110**(2):112-122

[44] Tanigaki K, Fujiura T, Akase A, Imagawa J. Cherry-harvesting robot. Computers and Electronics in Agriculture. 2008;**63**(1):65-72

[45] Van Henten EJ, Van Tuijl BV, Hemming J, Kornet JG, Bontsema J, Van Os EA. Field test of an autonomous cucumber picking robot. Biosystems Engineering. 2003;**86**(3):305-313

[46] Zanlorensi LA, Araújo VM, Guimarães AM. Automatic control and robotics for greenhouses: A review on heating technologies. Ibero|American Journal of Applied Computing. 2016;**4**(3):21-28

[47] Zhao Y, Gong L, Huang Y, Liu C. A review of key techniques of vision-based control for harvesting robot. Computers and Electronics in Agriculture. 2016;**127**:311-323

[48] Edan Y, Rogozin D, Flash T, Miles GE. Robotic melon harvesting. IEEE Transactions on Robotics and Automation. 2000;**16**(6):831-835

[49] Feng G, Qixin C, Masateru N. Fruit detachment and classification method for strawberry harvesting robot. International Journal of Advanced Robotic Systems. 2008;**5**(1):41-48

[50] Muscato G, Prestifilippo M, Abbate N, Rizzuto I. A prototype of an orange picking robot: Past history, the new robot and experimental results. Industrial Robot: An International Journal. 2005;**32**(2):128-138

[51] Plebe A, Grasso G. Localization of spherical fruits for robotic harvesting. Machine Vision and Applications. 2001;**13**(2):70-79

[52] Rath T, Kawollek M. Robotic harvesting of Gerbera Jamesonii based on detection and three-dimensional modeling of cut flower pedicels. Computers and Electronics in Agriculture. 2009;**66**(1):85-92

[53] Han KS, Kim SC, Lee YB, Kim SC, Im DH, Choi HK, Hwang H. Strawberry harvesting robot for bench-type cultivation. Journal of Biosystems Engineering. 2012;**37**(1):65-74

[54] Bechar A, Vigneault C. Agricultural robots for field operations. Part 2: Operations and systems. Biosystems Engineering. 2017;**153**:110-128

[55] Vasigh DJB. The Economic Impact of Unmanned Aircraft Systems Integration in the United States. Arlington, VA: Association for Unmanned Vehicle Systems International; 2013

[56] Mulla DJ. Twenty five years of remote sensing in precision agriculture: Key advances and remaining knowledge gaps. Biosystems Engineering. 2013;**114**(4):358-371

[57] Torres-Sánchez J, López-Granados F, Serrano N, Arquero O, Peña JM. High-throughput 3-D monitoring of agricultural-tree plantations with unmanned aerial vehicle (UAV) technology. PLoS One. 2015;**10**(6):e0130479

[58] Gago J, Douthe C, Coopman RE, Gallego PP, Ribas-Carbo M, Flexas J, Escalona J, Medrano H. UAVs challenge to assess water stress for sustainable agriculture. Agricultural Water Management. 2015;**153**:9-19

[59] Zarco-Tejada PJ, González-Dugo V, Berni JAJ. Fluorescence, temperature and narrow-band indices acquired from a UAV platform for water stress detection using a micro-hyperspectral imager and a thermal camera. Remote Sensing of Environment. 2012;**117**:322-337

[60] Bellvert J, Zarco-Tejada PJ, Girona J, Fereres E. Mapping crop water stress index in a pinot-noir vineyard: Comparing ground measurements with thermal remote sensing imagery from an unmanned aerial vehicle. Precision Agriculture. 2014;**15**(4):361-376

[61] Baluja J, Diago MP, Balda P, Zorer R, Meggio F, Morales F, Tardaguila J. Assessment of vineyard water status variability by thermal and multispectral imagery using an unmanned aerial vehicle (UAV). Irrigation Science. 2012;**30**(6):511-522

[62] Suárez L, Zarco-Tejada PJ, González-Dugo V, Berni JAJ, Sagardoy R, Morales F, Fereres E. Detecting water stress effects on fruit quality in orchards with time-series pri airborne imagery. Remote Sensing of Environment. 2010;**114**(2):286-298

[63] Bendig J, Bolten A, Bennertz S, Broscheit J, Eichfuss S, Bareth G. Estimating biomass of barley using crop surface models (csms) derived from uav-based rgb imaging. Remote Sensing. 2014;**6**(11):10395-10412

[64] Schirrmann M, Hamdorf A, Garz A, Ustyuzhanin A, Dammer KH. Estimating wheat biomass by combining image clustering with crop height. Computers and Electronics in Agriculture. 2016;**121**:374-384

[65] Duan S-B, Li Z-L, Wu H, Tang B-H, Ma L, Zhao E, Li C. Inversion of the prosail model to estimate leaf area index of maize, potato, and sun flower yields from unmanned aerial vehicle hyperspectral data. International Journal of Applied Earth Observation and Geoinformation. 2014;**26**:12-20

[66] Mathews AJ, Jensen JLR. Visualizing and quantifying vineyard canopy LAI using an unmanned aerial vehicle (UAV) collected high density structure from motion point cloud. Remote Sensing. 2013;**5**(5):2164-2183

[67] Verger A, Vigneau N, Chéron C, Gilliot JM, Comar A, Baret F. Green area index from an unmanned aerial system over wheat and rapeseed crops. Remote Sensing of Environment. 2014;**152**:654-664

[68] Garcia-Ruiz F, Sankaran S, Maja JM, Lee JS, Rasmussen J, Ehsani R. Comparison of two aerial imaging platforms for identification of huanglongbing-infected citrus trees. Computers and Electronics in Agriculture. 2013;**91**:106-115

[69] Sarkar SK, Das I, Ehsani R, Kumar V. Towards autonomous phytopathology: Outcomes and challenges of citrus greening disease detection through close-range remote sensing. In: 2016 IEEE International Conference on Robotics and Automation (ICRA); 16-21 May 2016; Stockholm, Sweden. IEEE; 2016. pp. 5143-5148

[70] Calderón R, Navas-Cortés JA, Lucena C, Zarco-Tejada PJ. High-resolution airborne hyperspectral and thermal imagery for early detection of verticillium wilt of olive using fluorescence, temperature and narrow-band spectral indices. Remote Sensing of Environment. 2013;**139**:231-245

[71] De Castro AI, Ehsani R, Poetz RC, Crane JH, Buchanon S. Detection of laurel wilt disease in avocado using low altitude aerial imaging. PLoS One. 2015;**10**(4):e0124642

[72] Aylor DE, Schmale DG, Shields EJ, Newcomb M, Nappo CJ. Tracking the potato late blight pathogen in the atmosphere using unmanned aerial vehicles and lagrangian modeling. Agricultural and Forest Meteorology. 2011;**151**(2):251-260

[73] Di Gennaro SF, Battiston E, Di Marco S, Facini O, Matese A, Nocentini M, Palliotti A, Mugnai L. Unmanned aerial vehicle (UAV)-based remote sensing to monitor grapevine leaf stripe disease within a vineyard affected by esca complex. Phytopathologia Mediterranea. 2016;**55**(2):262

[74] Giles DK. Use of remotely piloted aircraft for pesticide applications: Issues and outlook. Outlooks on Pest Management. 2016;**27**(5):213-216

[75] Faiçal BS, Costa FG, Pessin G, Ueyama J, Freitas H, Colombo A, Fini PH, Villas L, Osório FS, Vargas PA, et al. The use of unmanned aerial vehicles and wireless sensor networks for spraying pesticides. Journal of Systems Architecture. 2014;**60**(4):393-404

[76] Qin W-C, Qiu B-J, Xue X-Y, Chen C, Xu Z-F, Zhou Q-Q. Droplet deposition and control effect of insecticides sprayed with an unmanned aerial vehicle against plant hoppers. Crop Protection. 2016;**85**:79-88

[77] Antonelli MG, Auriti L, Beomonte Zobel P, Raparelli T. Development of a new harvesting module for saffron flower detachment. Romanian Review Precision Mechanics, Optics and Mechatronics. 2011;**39**:163-168

[78] Chatzimichali AP, Georgilas IP, Tourassis VD. Design of an advanced prototype robot for white asparagus harvesting. In: 2009 IEEE/ASME International Conference on Advanced Intelligent Mechatronics; 14-17 July 2009; Singapore. IEEE; 2009. pp. 887-892

[79] Aljanobi AA, Al-Hamed SA, Al-Suhaibani SA. A setup of mobile robotic unit for fruit harvesting. In: 19th International Workshop on Robotics in Alpe-Adria-Danube Region (RAAD 2010); 24-26 June 2010; Budapest, Hungary. IEEE; 2010

[80] Kitamura S, Oka K. Recognition and cutting system of sweet pepper for picking robot in greenhouse horticulture. In: IEEE International Conference Mechatronics and Automation; 29 July-1 August; Niagara Falls, Ont., Canada. IEEE; 2005. Vol. 4. pp. 1807-1812

[81] Qingchun F, Wengang Z, Quan Q, Kai J, Rui G. Study on strawberry robotic harvesting system. In: IEEE International Conference on Computer Science and Automation Engineering (CSAE), 2012; 25-27 May 2012; Zhangjiajie, China. IEEE; 2012. Vol. 1. pp. 320-324

[82] Sakai S, Iida M, Osuka K, Umeda M. Design and control of a heavy material handling manipulator for agricultural robots. Autonomous Robots. 2008;**25**(3):189-204

[83] Ceres R, Pons JL, Jimenez AR, Martin JM, Calderon L. Design and implementation of an aided fruit-harvesting robot (Agribot). Industrial Robot: An International Journal. 1998;**25**(5):337-346

[84] Hwang H, Kim SC. Development of multi-functional tele-operative modular robotic system for greenhouse watermelon. In: Proceedings of 2003 IEEE/ASME International Conference on Advanced Intelligent Mechatronics, 2003 (AIM 2003); 20-24 July 2003; Kobe, Japan, Japan. IEEE; 2003. Vol. 2. pp. 1344-1349

[85] Irie N, Taguchi N, Horie T, Ishimatsu T. Asparagus harvesting robot coordinated with 3-D vision sensor. In: IEEE International Conference on Industrial Technology, 2009. ICIT 2009; 10-13 Feb. 2009; Gippsland, VIC, Australia. IEEE; 2009. pp. 1-6

[86] Kondo N, Yamamoto K, Yata K, Kurita M. A machine vision for tomato cluster harvesting robot. In 2008 Providence, Rhode Island, 29 June-2 July 2008. American Society of Agricultural and Biological Engineers; 2008, p. 1

[87] Hayashi S, Ganno K, Ishii Y, Tanaka I. Robotic harvesting system for eggplants. Japan Agricultural Research Quarterly. 2002;**36**(3):163-168

[88] Liu TH, Zeng XR, Ke ZH. Design and prototyping a harvester for litchi picking. In: International Conference on Intelligent Computation Technology and Automation (ICICTA), 2011; 28-29 March 2011; Shenzhen, Guangdong, China. IEEE; 2011. Vol. 2, pp. 39-42

[89] Ismail W, Ishak W, Kit WH, Awal M. Design and development of eggplant harvester for gantry system. Pertanika Journal of Science & Technology. 2010;**18**(2):231-242

[90] Arima S, Kondo N. Cucumber harvesting robot and plant training system. Journal of Robotics and Mechatronics. 1999;**11**:208-212

[91] Bonadies S, Lefcourt A, Gadsden SA. A survey of unmanned ground vehicles with applications to agricultural and environmental sensing. In: SPIE Commercial+ Scientific Sensing and Imaging. International Society for Optics and Photonics; 2016. pp. 98660Q-98660Q

[92] Bechar A, Vigneault C. Agricultural robots for field operations: Concepts and components. Biosystems Engineering. 2016;**149**:94-111

[93] Zhan Q, Cai Y, Liu Z. Near-optimal trajectory planning of a spherical mobile robot for environment exploration. IEEE Conference on Robotics, Automation and Mechatronics; 21-24 Sept. 2008; Chengdu, China. IEEE; 2008. pp. 84-89

[94] Hernández JD, Sanz D, Rodríguez-Canosa GR, Barrientos J, del Cerro J, Barrientos A. Sensorized robotic sphere for large exterior critical infrastructures supervision. Journal of Applied Remote Sensing. 2013;**7**(1):073522-073522

[95] Hernández JD, Barrientos J, del Cerro J, Barrientos A, Sanz D. Moisture measurement in crops using spherical robots. Industrial Robot: An International Journal. 2013;**40**(1):59-66

[96] Dorhout D. Prospero, The Robot Farmer. Available from: http://www.dorhoutrd.com/prospero_robot_farmer [Accessed: January 2017]

[97] Wood R. RoboBees Project. Harvard University; Cambridge, Massachusetts, United States; 2015

[98] Avellar GS, Pereira GA, Pimenta LC, Iscold P. Multi-UAV routing for area coverage and remote sensing with minimum time. Sensors. 2015;**15**(11):27783-27803

[99] Chao H, Baumann M, Jensen A, Chen Y, Cao Y, Ren W, McKee M. Band-reconfigurable multi-UAV-based cooperative remote sensing for real-time water management and distributed irrigation control. IFAC Proceedings Volumes. 2008;**41**(2):11744-11749

[100] Pesticide Sales Statistics, by Major Groups in 2014. Eurostat Statistics Explained. Available from: http://ec.europa.eu/eurostat/statistics-explained/index.php/Pesticide_sales_ statistics [Accessed: December, 2016]

[101] Pimentel D. Environmental and economic costs of the application of pesticides primarily in the United States. Environment, Development and Sustainability. 2005;**7**:229-252

[102] Del Cerro J, Barrientos A, Sanz D, Valente J. Aerial fleet in RHEA project: A high vantage point contributions to ROBOT 2013. In: ROBOT2013: First Iberian Robotics Conference. Springer International Publishing; 2014. pp. 457-468

[103] Rabatel G, Labbé S. Registration of visible and near infrared unmanned aerial vehicle images based on Fourier-Mellin transform. Precision Agriculture, 2016, vol. 17, no 5, p. 564-587

[104] López-Granados F. Weed detection for site-specific weed management: Mapping and real-time approaches. Weed Research. 2011;**51**(1):1-11

[105] Conesa-Muñoz J, Ribeiro A, Andujar D, Fernandez-Quintanilla C, Dorado J. Multipath planning based on a NSGA-II for a fleet of robots to work on agricultural tasks. In: Abbass H, Essam D, Sarker R, editors. IEEE World Congress on Computational Intelligence. Red Hook, NY: Congress of Evolutionary Computation (CEC), Curran Associates Inc; 2012. pp. 2236-2243

[106] Romeo J, Guerrero JM, Montalvo M, Emmi L, Guijarro M, Gonzalez-de-Santos P, et al. Camera sensor arrangement for crop/weeds detection accuracy in agronomic images. Sensors. 2013;**13**:4348-4366

[107] Xiaoyuan Y, Jiwei D, Tianjie Y, Qingfu Q. A method for improving detection of gas concentrations using quadrotor. In: Information Technology, Networking, Electronic and Automation Control Conference; 20-22 May 2016; Chongqing, China. IEEE; 2016. pp. 971-975

Chapter 5

Service Robots for Motion and Special Applications on the Vertical Oriented Walls

Marcel Horák, František Novotný and Michal Starý

Additional information is available at the end of the chapter

http://dx.doi.org/10.5772/intechopen.70037

Abstract

This chapter is focused on the area of mobile systems of service robots for motion on the vertically oriented glass walls (e.g., facades of high-rise building) with the aim of their using in many inspection and technological applications. Preliminary part clearly maps the basic mechanical principles and approaches to mobile platform design with respect to the concept of kinematic chain and type of actuators. Conclusions of extensive research activities are presented, and on this background, the new design development of mechanics of robot mobile platform was made and uses two parallel placed parallelograms. The control system is based on an industrial computer, includes a module for wireless communication, and is equipped with a laser and an ultrasonic position sensor. Movement members are equipped with individual electric actuators and vacuum gripping system, which consists of smart ejectors in combination with active suction cups. Given that the load character of the suction cups during the robots movement on the vertical wall is very unfavorable, considerable authors' attention has been paid to the analysis of the deformation behavior of suction cups so as to determine the limits of external radial load to the stable contact, and discusses the possibility of increasing the radial load of gripping elements in relation to the contact surfaces character and vacuum levels.

Keywords: service robot, vertical wall, vacuum, servo actuator, special service application, combined vacuum element, suction cup, safety coefficient, numerical simulation

1. Introduction

Construction of high-rise buildings with facades made of protective glass fixed to supporting grids is a contemporary trend in modern architecture. There has been a long-term increase in the demand for devices and new technologies, enabling users to deal effectively with problems concerning cleaning, inspection, installation and other service applications. Inspection of the potential surface disruptions of large pressure vessels is just one of many examples. A number of service robots or, more precisely, mobile platforms with different movement characteristics as well as various abilities to deal with surface height differences have been developed recently.

The existing mobile platforms for motion on vertical glass walls have had either the stepping [1–4], continuous [5, 6] or pseudo-continuous motion [7], and the force of holding on the glass wall has been produced using the active-vacuum suction cups and modules (holding systems using materials with a high degree of adhesion are also currently presented). The stepping motion of the mobile platform is realized by means of rectilinear pneumatic motors or reciprocating linear electric actuators. Vacuum in the leg suction cups is released after platform body is fixed to a wall using the force produced in the active suction cups; then, the legs are pulled away from the contact with the glass wall and repositioned by stepping translational motion.

Furthermore, the legs are pulled back to the glass wall and fixed by vacuum set off in leg suction cups. Body of a mobile platform performs similar cycle of the motion by step. Among most significant disadvantages of this design belong to a complicated design of the electric-actuator transformational unit, a considerable weight of translational motion units and an intermittent motion of the mobile platform.

Approaches based on a design with the pseudo-continuous motion are realized by rotary motion of the stepping unit with a horizontal or parallel movement perpendicularly to the glass wall plane. Continuous systems are based on use of oval suction cups — vacuum chambers located on conveyor elements. Designs using collapsible elements on rotary wheels are quite exceptional. These approaches involve using materials coating with adhesive layer [8].

On the basis of theoretical studies, the first generation of the service-robot mobile platform [9–11] was developed and its operation was successfully tested on a glass wall. Verified design of the motion principle was subsequently optimized and used for the second generation of the mobile platform [12]. The motion was not changed as a matter of principle, but by application of the new frame design and smart servo-actuators. That brought required reduction of weight and effective carrying capacity was increased up to ca 25 kg. Moreover, motion possibilities were extended by a next motional axis, allowing the platform to be rotated around its axis perpendicularly to the contact wall plane. This offers an advantage in compensation of motional deviations from the preselected direction, for changing of the platform direction from vertical to horizontal and so on.

Optimized motion platform with improved load capacity might be used for many technological operations. Currently, two superstructures were realized: one for dry or wet cleaning of the glass facades of high-rise buildings and the other for camera-assisted inspection tasks.

Load capacity is for mobile platforms essential and depends on the mechanical and electrical design of the platform. Therefore, new composite materials and design techniques are used, the platform weight is reduced and load capacity is increased. Exceptional demand has been put on a design of new gripping system based on combined vacuum elements, when materials with a high degree of surface adhesion [13] have been applied. Authors describe this in detail at the end of this chapter.

There is the initial research and first generation of service robots for movement on vertical glass facades described in the second chapter. The third chapter deals with optimized version — second generation of the service robot. Safety aspects of the vacuum suction system are described more in detail in Chapter 4. Chapter 5 introduces possibilities for radial carrying capacity increase. The use of deformation passive suction cups in climbing service robotics is discussed in the last section.

2. Development of the first service robot generation

The service robot mobile has been developing since 2008. During this time period, the platform has gone through several stages of development. The initial stage was represented by a project of the holding mechanical concept, and another one included drives and control design; then, subsequently, a remote communication including a graphical user interface was solved. Several robot versions (**Figure 1**) were designed during the project. They differed especially in maneuverability, that is, possibility to change the robot motional and orientation sequence in dependence on the surrounding environment topology. Patented pseudo-continual stepping motion system was used in all of the versions. This is presented by a cyclic alternative movement of the robot legs and body. The step size is firmly fixed by an interconnecting crank length.

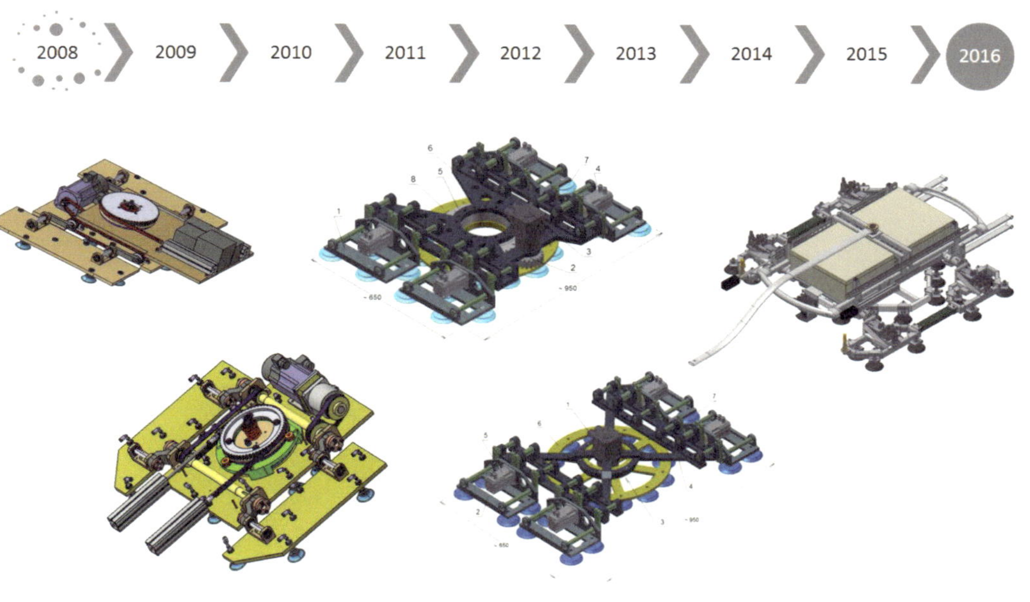

Figure 1. Development of the mobile platform in variants.

The first version of a functional sample of the robot platform (**Figure 2**) was laboratory-tested with very good results. It is a partially autonomous solution with a stepped motion during which the platform legs and body motions take place step by step. A force bond, that is, keeping the platform on the vertical wall, is provided by active suction cups in combination with a system of smart pneumatics. A pneumatic circuit was projected with respect to safety aspects of the motion along vertical walls. An arrangement of 24 suction cups was optimized in view of the model results when the specific load of suction pads and the vacuum interruption during the platform motion over the real construction of the building glass lining were indicated. Suction cups were divided up four sections at six pieces each. Each section is controlled by one smart ejector having the vacuum indication and quick response. Each suction cup is equipped with a self-closing vacuum supply valve if the vacuum space tightness under the suction cup is disrupted.

A drive is formed by a pair of parallelograms, belt gear and three-phase step motor which is controlled by a power unit FM STEPDRIVE directly connected with a control unit based on a program logic control [14, 15]. This concept can be considered to be promising for the cleaning robot application intended, that is, when the robot will use for a building with its skin negative inclination (an inverted pyramid system). In the course of the driving crank rotation, the pair of the platform "feet" and "body" is set in motion alternately and stepping occurs when connecting vacuum into appropriate sections of the suction cups of motional elements. Quick alternations of an under pressure in the vacuum suction cups and their reaeration are provided in the connection with measurements of the crank positions and the vacuum values. The motional sequence control is allowed applying Programmable Logic Controller (PLC).

Main functional units were verified during tests of the first version of the robotic mobile platform. The accent was put on the proper mechanics of the chassis, sensory subsystem and driving system and control. Unequivocal system defects and limits were established after subsequent interpretation of operating data. Based on the obtained experience, demands for the new generation service robot were defined. Situation is clearly visible in **Figure 3**.

Figure 2. The first generation of the mobile platform.

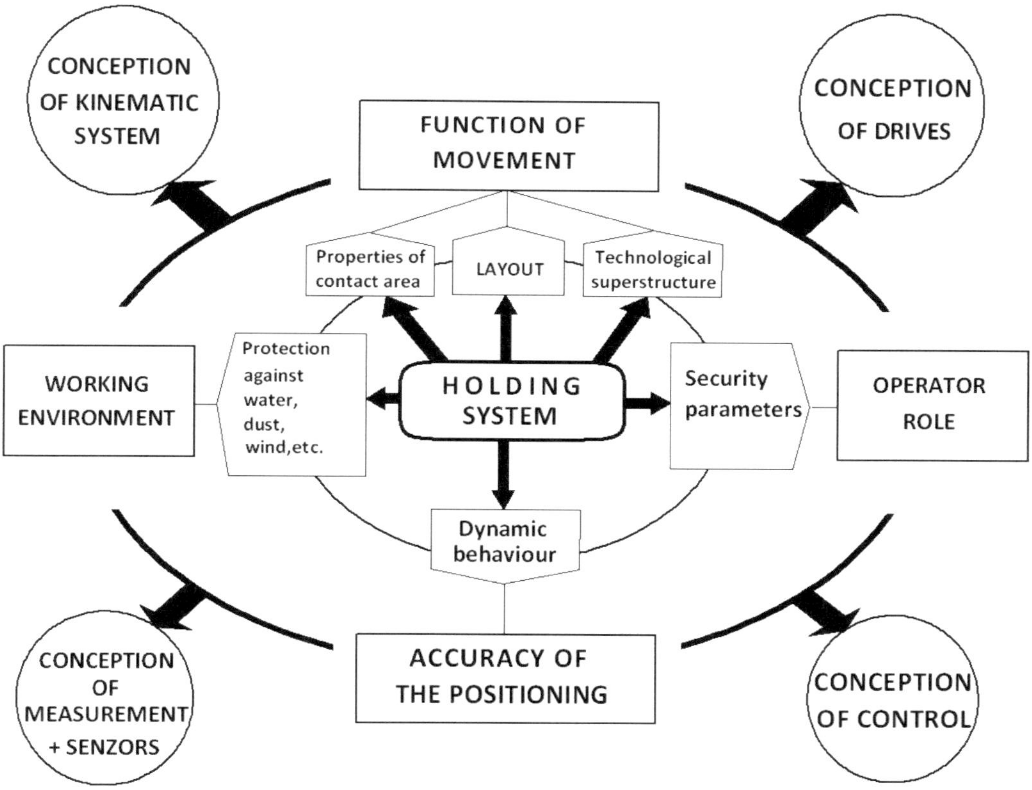

Figure 3. Overview of system links of the mobile platform.

Optimal functioning of the robot is obviously significantly affected by a design of its kinematical chain, drives and holding system. It is appropriately a determining factor for fixing safety parameters; these parameters influence the robot dynamic behavior, layout, contact plane characteristics and eventually staff operation.

3. Second generation of the service robot

Based on the laboratory tests of function sets of the first-generation mobile platform of the robot as well as subsequently found defaults, system connections (relations) and demands for a design of the robot second generation were defined unambiguously with the aim of improving the system (drives) dynamic behavior, maneuverability, safety, control comfort, progress of use properties, effective weight and last but not least also cover design.

The second-generation platform (**Figure 4**) comprises a compact duralumin frame interconnected with a pivoted chassis through a rotary servo drive; by this way, the robot body is created. Moreover, the pivoted chassis is fitted with hinged active suction cups the arrangement of which can be set up due to a character of the contact plane (façade system) geometry. The platform

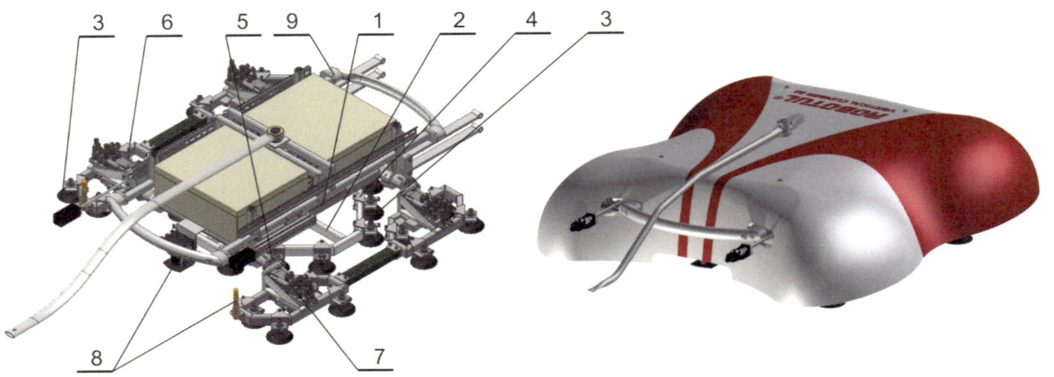

Figure 4. Computer model of the second generation of the mobile platform (*1* — frame, *2* — pivoted chassis, *3* — suction cup, *4* — adjustable holder, *5* — crank, *6* — motional leg, *7* — rotary servo drive, *8* — sensors, *9* — extended frame for superstructures installation).

frame is connected with four legs by means of pivoted cranks. Each leg is fitted with suction cups and an individual electric rotary servo drive [16].

Robotic legs are joined together by a flexible part on both sides of the platform. This flexible part works as a parallelogram in its motional function, and simultaneously, it compensates the leg positional inaccuracies in the contact with a glass wall or, for example, low-profile construction elements of a façade anchoring systems.

The motional sequence is realized as follows: under pressure is brought into the suction cups, and the pivoted body is fixed on the vertical wall. During the initial fixation, leg cranks are quarter-turned normally to the contact surface. This allows turning the platform frame to a required direction. Subsequently, the legs are turned back by ca. 90° and fixed on the wall by bringing a vacuum in leg suction cups.

The vacuum under the body suction cups was cancelled in the second phase. The cranks are turned by 180° with legs fixed on the wall; thus, the body of the service robot is moved one step forward. Continual forward or backward motion is realized by alternating motion of the platform body and legs, and cranks are always rotated by ca. 180°. Appropriate pressure characteristics are synchronized with the movement. To achieve the desired operating safety, suction cup system was divided into several independent sections with the local source of vacuum; contemporary ejectors with integrated sensors of vacuum and a blowing pulse function were used.

For safety reasons, an extension frame was designed, allowing a pivoted suspension arm, power chains and the robot composite cover to be attached (**Figure 5**). Construction elements for fixing technological extensions and operating grab handles are parts of the frame. In **Figure 6** is illustrated the service robot final solution with a dock station, allowing the system to be provided with a user Human Machine Interface (HMI), contact panel making remote control, standby battery supply, computer and so on. From the standpoint of the robot control, a user program was created based on the platform of the programming language ANSI C, respectively C++, with a support of software tool Automation Studio from the B&R Company. The program is divided up into several independent logic structures for which partial function

Figure 5. The mobile platform without and with the composite cover.

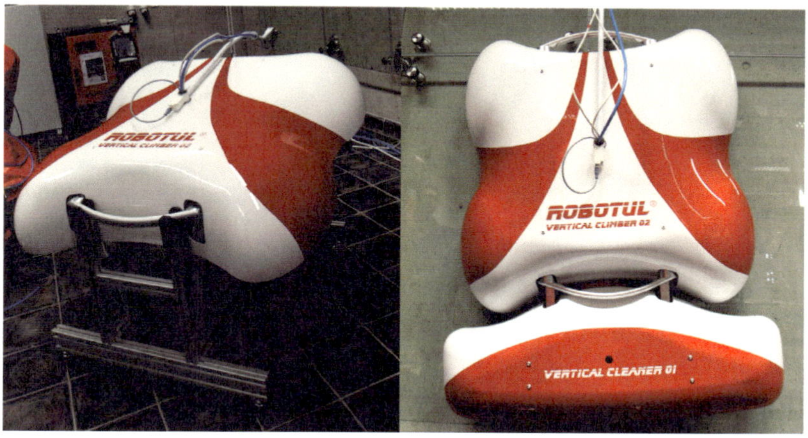

Figure 6. Service robot with a dock station and cleaning unit.

blocks were created, making a communication of the industrial PC with digital servo amplifiers of the rotary drives possible.

In view of the carrying capacity, respectively, the platform weight minimization, the frame (**Figure 4**, pos. 1) strength was analyzed [17]. Based on this analysis, a minimum thickness of the frame bearing profiles wall and connecting plate was predicted, considering tilting moment 40 Nm in the presumptive mass center distance 100 mm and the total weight of 41 kg. Results from the computer simulation propose a weldment arrangement (**Figures 7** and **8**).

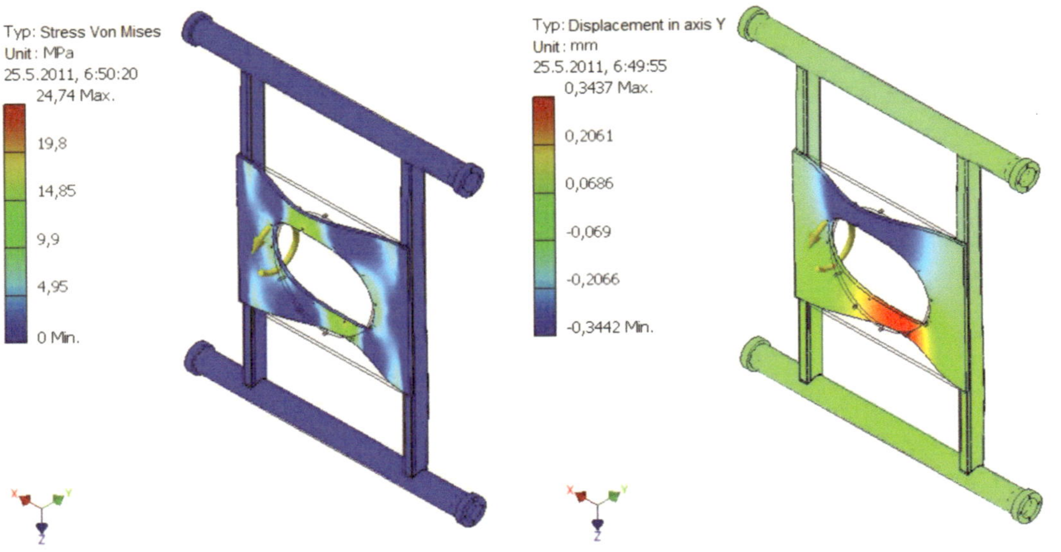

Figure 7. Stress and deformation fields in the frame without ribs having the connecting plate of 6 mm thickness.

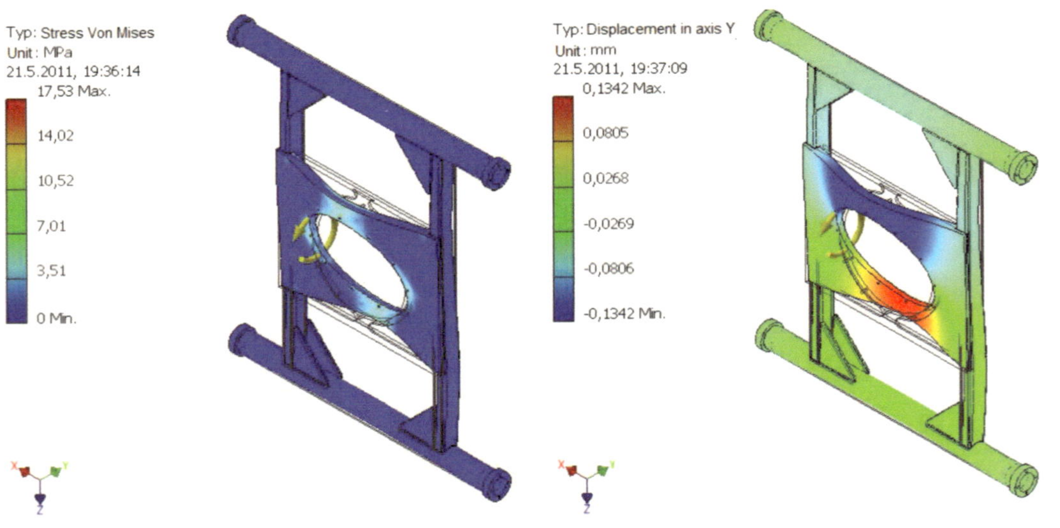

Figure 8. Stress and deformation fields in the optimized frame.

Stress and deformation fields in the frame were analyzed based on the obtained data. It allowed adapting the initially proposed solution by the use of additional reinforcing ribs. These were placed inside the frame, and further, they will be used for placing other accessories of the mobile platform in common with a free space outside the frame. The thickness of the plate was increased from 6 to 15 mm in the place of ribs to reach higher rigidity. It also reduced stress in the optimized frame by 29% — 17.53 MPa in comparison with the initial design. The maximum deformation was reduced by ca 60% — 0.134 mm.

4. Safety aspects of vacuum holding system of robot

Authors' considerable attention was directed to problems being connected with designing a vacuum gripping system influencing decisively the robot stability and safety holding on vertically oriented walls. For that purpose a computer analysis of deformation behavior of the individual suction cup under dominant loading in the radial direction was created. On the base of obtained results a stability limit was defined step by step for which a proportion between an absorbed elastic energy U_{el} and energy of adhesion U_{ad} is very important from the mathematical standpoint [18, 19]. If we proceed from the assumption that

$$U_{el} = E\lambda h^2 \text{ and } U_{ad} = -\Delta\gamma\lambda^2, \tag{1}$$

then it is true

$$\theta = \frac{U_{el}}{U_{ad}} \approx \frac{Eh^2}{\Delta\gamma\lambda}, \tag{2}$$

where E is modulus of contact elastomer elasticity, λ is width of contact body, h is roughness and $\Delta\gamma$ is change in free energy per unit area in which the elastomer is in the contact with the body (surface roughness are filled with elastomer).

If $\theta << 1$, that is an ideal contact when contact surfaces of bodies copy each other perfectly. If it is true that $\theta >> 1$, then the contact is only between maxima of the surface unevenness.

Alternatively to the contact analytical description, a numerical model of the suction cup and the glass contact surface was made up so that the real behavior of elastomer was replaced by the Mooney-Rivlin rheological model [20, 21], and appropriate material constants were computed from the relation (for only uniaxial tension)

$$F = 2A_0 \cdot \left(1 - \frac{1}{\lambda_1^3}\right) \cdot (\lambda_1 c_{10} + c_{01}), \tag{3}$$

where F is tension force, A_0 is an initial cross-sectional area of the test sample, λ_1 is a deformation and c_{10} and c_{01} are Mooney-Rivlin material constants.

On the basis of the calculation using the finite-element method, three principal limit states can be defined (**Figure 9**) for the selected type of the active suction cup of the holding system which correspond to the suction cup contact profile depending on the level of external radial force according to the force analysis carried out in detail in Ref. [22].

If equilibrium conditions of an interaction between the suction cup and contact surface are analyzed in detail [23], then it is obviously true that

$$T + \Delta T_P - F_{RAD} = 0, \tag{4}$$

$$F_U - F_B - F_{AX} - \Delta F = 0. \tag{5}$$

Along the suction cup axis, external axial forces F_{AX} and F_G are acting induced by the under pressure applying on the suction cup active surface. As a result of the suction cup thrust, a

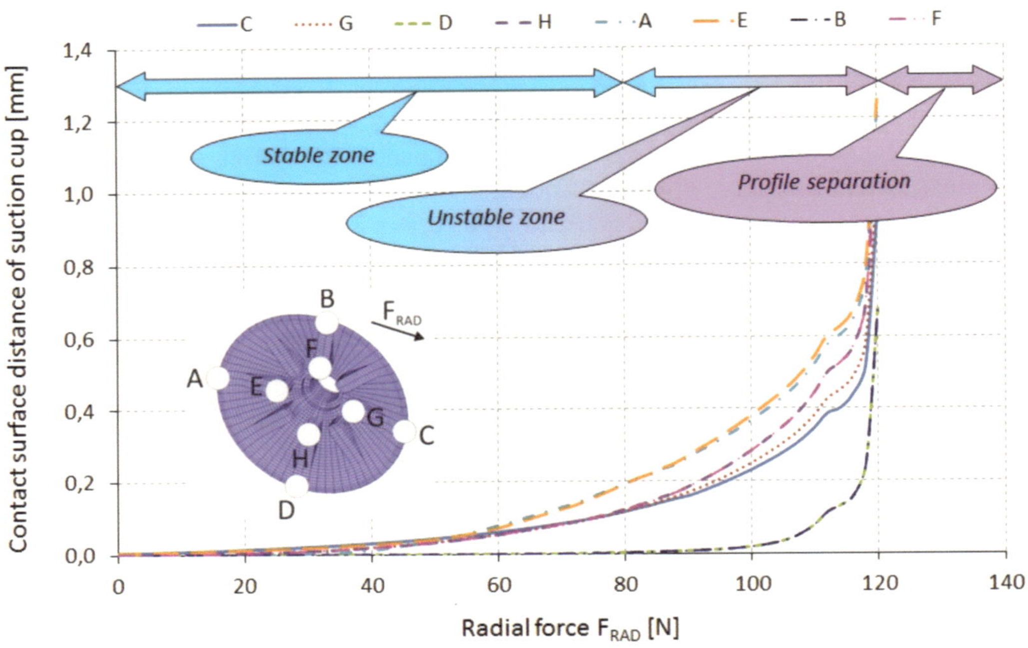

Figure 9. Typical progression of suction cup deformation in the radial direction.

force F_B is acting further, and between the suction cup bottom and the contact wall, an additional force ΔF is making evident. Then, a radial force F_{RAD} at the plane of gripping and a displacing force T on the edge sealing surface are acting. Owing to a friction on the suction cup bottom caused by the thrust normal force ΔF, it is possible to define an additional friction force $\Delta T_p = \Delta F.\ \mu$ (μ is friction coefficient). A theoretical gripping force F_{Gtheor} can be defined for the state of limit equilibrium, that is, when the force ΔF will equal zero.

$$F_{Gtheor} = \frac{F_{RAD}}{\mu} + F_{AX} + F_B. \tag{6}$$

It is not easy to determine the force F_B, and particular calculations were necessary to base on the data found experimentally. If material properties of the suction cup as well as boundary conditions are known, it is possible to use the rule of thumb using the finite-element method. In many cases, the presumption is that $F_B \ll F_{AX}$ and it can be ignored. As for the real gripping force F_G, the relation is generally true (only for centric holding)

$$F_G = k' \cdot \frac{F_{RAD}}{\mu} + k'' \cdot F_{AX}, \tag{7}$$

where k' (safety against sliding) and k'' (safety against breakoff) are appropriate safety factors corresponding to the direction of the suction cup greatest load for which it is true that $k' > k''$. Usual factor values are $k'' = 6–8$ and $k'' = 4–5$.

http://dx.doi.org/10.5772/intechopen.70037

5. Possibilities for an increase of the suction cup radial carrying capacity

With the aim of decreasing the robot demands of energy, a new concept was solved relevant to a so-called combined gripping element inducing gripping force by means of vacuum and adhesion. The presumption was that higher level of a possible safety load along the radial direction, that is, along the contact plane, would be reached, and at the same time, the consumption of pressure air would be decreased or kept at least.

The principle of such solution [24] is represented in **Figure 10**. It lies in that: when the gripping element *1* comes into contact with the object handled *7*, the created space *10, 11* being sealed with the sealing rim *2* is evacuated through the hole *8*; at the same time, the bearing plate *4* is moved optimally into contact with the object *7* by means of the piston *3* which is connected with the bearing plate *4,* and simultaneously, it sealed the space *10*. Part and parcel of the bearing plate *4* is the adhesive layer *5,* forming the contact interface *12* with the object *7*. A thrust level between the plate *4*, respectively layer *5*, and the object *7* is possible to control by the value of pressure in the regulative space *9* above the piston *3* so that the ideal level of thrust can be reached, based on the rim *2* deformation, mechanical properties as well as a surface profile of the object *7*. In this way, it is possible to provide maximal growth of friction forces during holding.

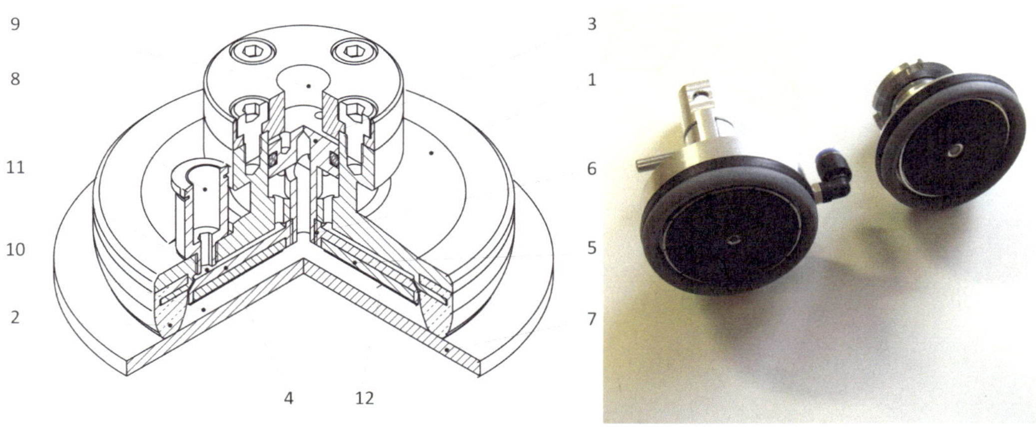

Figure 10. Combined vacuum gripping element.

6. The use of deformation passive suction cup in a holding system

An optimized and simultaneously original solution of the holding system is a possibility to use only deformation, so-called passive suction cups. This solution would allow the service robot to be operated without necessity to use an active source of vacuum dependent on pressure air [25]. Based on equilibrium of forces, a gripping force can be determined provided that the suction cup diameter is not changed according to the relation

$$F_G = \varsigma \cdot F_{Gtheor} = \frac{\pi \cdot d^2}{4} \cdot (1 - K) \cdot p_a \cdot \varsigma, \tag{8}$$

where ζ is a correction to the suction cup rigidity that is introduced due to the real situation where in the moment of pressing the suction cups into contact are in a closed volume pressure less than atmospheric pa. ζ acquires values in the interval from 0.6 to 0.8, and K is a volume ratio which lies in the interval from 0.2 to 0.5 for typical suction cups.

When applying in real conditions, it is necessary to realize that the gripping force values are influenced by the surface quality, material properties of the suction cup, as well as they depend on the holding time, which is connected with leakages of the closed volume between the suction cup and object. For this reason, it is necessary to choose relatively high values of the safety factor $k = 4$–6.

A suitable option for using passive suction cups shows a solution when the suction cup will be deformed actively, so that the under-pressure level can be regulated on the one hand, and leakage losses can be possibly compensated for further growth of deformation on the other hand.

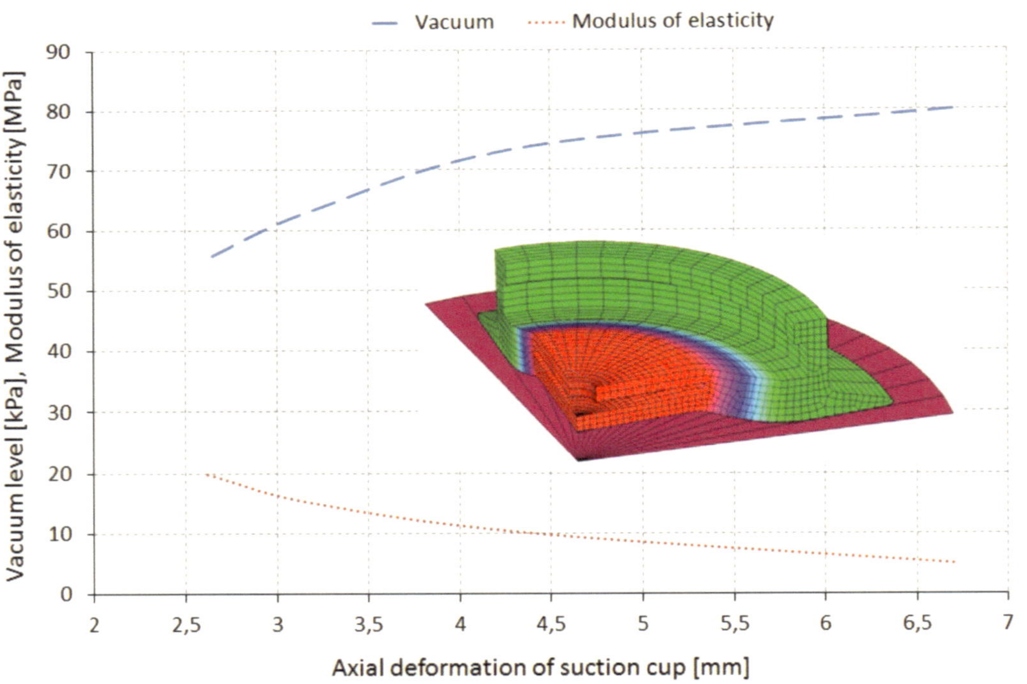

Figure 11. Axial deformation of passive suction cup.

http://dx.doi.org/10.5772/intechopen.70037

However, there is one great problem, namely that depending on material, geometrical properties of the suction cup and a vacuum final value, this solution needs applying a relatively considerable active force which will induce an appropriate deformation.

The diagram in **Figure 11** shows a course of the suction cup deformation depending on the obtained values of vacuum as well as the suction cup material properties given by a modulus of elasticity E at axis loading with the active force $F_A = 100$ N. It is evident from the given diagram that for $E = 10$ MPa ($c_{01} = 0.3$ MPa, $c_{10} = 1.3$ MPa) the available theoretical level of vacuum is *ca.* 70 kPa. It is necessary to realize that due to a great number of suction cups creating the robot holding system, the given solution is not realizable without a specific technical solution of the mechanical transforming block of the suction cup because the final force load of an additional drive, allowing the suction cups to be controlled directly, will be too high.

7. Conclusion

Basic mechanical systems of the service robots were described, and **concrete solution of the robot mobile platform was presented** in this chapter. The platform allows realizing a motion on vertically oriented smooth-surfaced walls. This solution allows special technological and inspection extents to be installed with a view to realizing, for example, cleaning, assembling and detecting processes. Moreover, a methodology solving deformation contact tasks was managed inclusive of introducing specific boundary conditions being connected with the pressure definition on the contact interface. Also, a deformation analysis of the radially loaded suction cup of the robot holding system was carried out.

Further problems of **defining theoretical and real gripping force of the individual suction cup during axial as well as radial loading were discussed**, and adequate safety factors corresponding to a character of load and operational conditions were determined.

A combined vacuum gripping element inducing gripping force on the principle of vacuum and adhesion **was presented,** having an evident influence on the contact stability. Based on laboratory tests, it was shown that it increases the radial carrying capacity against the system without any adhesion layer by a few 10%.

In conclusion, a **possibility to use only deformation suction cups** in holding systems of mobile robots was analyzed as an alternative to the standard-chosen solution with active suction cups with the aim of eliminating the source of pressure air and vacuum circuit of the mobile platform.

Acknowledgements

The results of this project LO1201 were obtained with co-funding from the Ministry of Education, Youth and Sports as part of the targeted support from the program entitled "Národní program udržitelnosti I."

Author details

Marcel Horák*, František Novotný and Michal Starý

*Address all correspondence to: marcel.horak@tul.cz

Department of Mechatronic Systems, Institute for Nanomaterials, Advanced Technologies and Innovation, Technical University of Liberec, Liberec, Czech Republic

References

[1] Lunch GA, Clark JE, Lin PC, Koditschek DE. A bioinspired dynamical vertical climbing robot. The International Journal of Robotics Research. 2012;**31**:974-996

[2] Alexandrescu N, Apostolescu TC, Duminica D, Udrea C, Ionascu G, Bogatu L. Construction of a vertical displacement service robot with vacuum cups. In: Gacovski Z, editor. Mobile Robots — Current Trends. Croatia: InTech; 2011. pp. 215-238

[3] Albagul A, Asseni A, Khalifa O. Wall climbing robot: Mechanical design and implementation. In: Proceedings of Recent Advances in Circuits, Systems, Signal and Telecommunications. Puerto Morelos, Mexico: WSEAS Press; 2011. pp. 28-32. ISBN: 978-960-474-271-4

[4] Luk BL, Liu KP, Collie AA, Cooke DS, Chen S. Tele-operated climbing and mobile service robots for remote inspection and maintenance in nuclear industry. Industrial Robot: An International Journal. 2006;**33**(3):194-204

[5] Silva MF, Barbosa RS, Oliveira AL. Climbing robot for ferromagnetic surfaces with dynamic adjustment of the adhesion system. Journal of Robotics. **2012**:16. DOI: 10.1155/2012/906545

[6] De Fazio TL. Wall Climbing Robot. US Patent No. 6793026 B1. September 21, 2004

[7] Xiao J, Morris W, Chakravarthy N, Calle A. City-climber: A new generation of mobile robot with wall-climbing capability. In: Proceedings of SPIE. Vol. 6230; 2006 SPIE Defense & Security Symposium, Orlando, Florida, USA: The International Society for Optical Engineering; 2006

[8] Kim H, Kim D, Kim HY, Lee K. Development of a wall-climbing robot using a tracked wheel mechanism. Journal of Mechanical Science and Technology. 2008;**22**:1490-1498. DOI: 10.1007/s12206-008-0413-x

[9] Novotný F, Horák M, Obst M. Design study of service robot chassis for motion on glass vertical wall. In: 8th International Carpathian Control Conference. Košice, Slovakia: Technical University of Košice; 2007. pp. 492-495. ISBN: 978-80-8073-805-1

[10] Horák M, Novotný F. Service robots for building glass facades cleaning. Acta Mechanica Slovaca. 2008;**12**(2-A):273-276. ISSN: 1335-2393

[11] Novotný F, Horák M. Computer modelling of suction cups used for window cleaning robot and automatic handling of glass sheets. MM Science Journal. 2009;**2**:113-116. ISSN: 1803-1269

[12] Novotný F, Horák M, Plavec M. Mobile Platform for Movement on Vertical Wall. CZ Patent No. 303697, 2013

[13] Horák M, Novotný F. Special gripping elements for handling with flat objects. Applied Mechanics and Materials. 2013;**282**:27-32. DOI: 10.4028/www.scientific.net/AMM.1000.27. ISSN: 1660-9336

[14] Novotný F, Horák M. Autonomous robotic mobile platform. In: XVI International Scientific Symposium TU Liberec-TU Dresden; Liberec, Czech Republic. Liberec: Technical University of Liberec; 2007. ISBN: 978-80-7372-247-0

[15] Machytka J, Horák M, Novotný F. Control conception of the service robot mobile platform. In: 5th International PhD Conference on Mechanical Engineering — PhD 2007. Pilsen, Czech Republic: University of West Bohemia; 2007. pp. 145-148. ISBN: 978-80-7043-597-7

[16] Horák M, Novotný F, Starý M, Černohorský J. New generation of mobile platform of service robot for motion along vertical walls. Applied Mechanics and Materials. 2014; **613**:126-131. DOI: 10.4028/www.scientific.net/AMM.613.123. ISSN: 1660-9336

[17] Novotný F, Horák M, Plavec M. Service robot for motion on the vertical oriented walls. Applied Mechanics and Materials. 2013;**282**:39-44. DOI: 10.4028/www.scientific.net/AMM.1000.39. ISSN: 1660-9336

[18] Persson BNJ, Bucher F, Chiaia B. Elastic contact between randomly rough surfaces: Comparison of theory with numerical results. Physical review B. 2002;**65**(184106):7

[19] Zilberman S, Persson BNJ. Adhesion between elastic bodies with rough surfaces. Solid State Communications. 2002;**123**:173-177

[20] Mooney MA. Theory of large elastic deformations. Journal of Applied Physics. 1940;**11**:582-592. DOI: 10.1063/1.1712836

[21] MSC Software Corporation. MSC. Marc Volume A: Theory and User Information. Santa Ana, CA: MSC.Software Corporation; 2014

[22] Novotný F, Horák M. Problems of loading capacity of suction cups in the radial direction. In: OPTIROB 2010; 28–30 May. Calimanesti, Romania: Research Publishing Services, Singapore; 2010. pp. 70-74. ISBN: 978-981-08-5840-7

[23] Horák M. Dynamic Handling with Flat Glass. 1st ed. Liberec, Czech Republic: Technical University of Liberec; 2008. p. 165. ISBN: 978-80-7372-425-2

[24] Horák M, Novotný F. Increases in the radial capacity of vacuum gripping elements. MM Science Journal. October 2011;**Special Edition**:122-127. ISSN: 1803-1269

[25] Brockmann W, Mösch F. Climbing without a vacuum pump. In: Proceedings of the 7th International Conference CLAWAR 2004. Berlin, Heidelberg: Springer; 2005. pp. 935-942. DOI: 10.1007/3-540-29461-9_91. ISBN: 978-3-540-22992-6

Chapter 6

Robot Engraving Services in Industry

Silvia Anton, Florin Daniel Anton and
Mihai Constantinescu

Additional information is available at the end of the chapter

http://dx.doi.org/10.5772/intechopen.69928

Abstract

This chapter presents a software system created to allow to design complex robot trajectories in the cartesian space, used for tasks like milling or engraving with the aid of a robot. The experimental setup was created using an ABB (ASEA Brown Boveri) IRB 140 robot on which was mounted an engraving tool (which was equipped with a high frequency vibrating pin) or a drilling tool (an electric milling machine controlled with I/O signals) and a computer which is executing a CAD application allowing the design of the path that the robot must follow in order to execute the engraving/milling task. The robot job or task is defined by the set of surfaces of the raw object and of the object after being processed, by subtracting these two surfaces the robot task is given by the set of points resulted after the subtraction. The subtraction operation offers a set of points which must be connected in order to obtain the final trajectory, if the task is to engrave a shape then the tool will be normal to the object surface. The algorithm which is generating the surface is presented along with some experimental results.

Keywords: surface extraction, trajectory definition, task modelling, engraving services, industrial robot

1. Introduction

In the last years the manufacturing has changed from mass production to batch production in which a medium or small volume of products are created, the manufacturing process is executed on demand, batches being executed differently based on the client specifications.

Nowadays production lines must be flexible and reconfigurable, changing from a version of a product to another sometimes requires changes in the manufacturing line which higher the costs, because of that there is the trend to use industrial robots which are able to adapt to these changes only by changing the program. The robots are programmed to execute not only a task, but multiple tasks in order to lower the costs for creating the products, these additional tasks executed by the robots can eliminate other expensive machines such engraving or CNC (computer numerical control) milling machines [1, 2]. In comparison, a robot is much cheaper than a CNC milling machine, but is able to complete milling tasks, with acceptable performances [3, 4].

The main goal of the project that the which is presented in this chapter presents ishas the main goal to developcreate a system which allows to industrial robots to will offer the possibility to design and describe complex surfaces. The system allows to industrial robots to generate complex surfaces which then an industrial robot will follow for executing milling or engraving tasks. The surface will be followed by using programs where the robots will execute procedural motions of the tool, this allows, allowing to processcreate customized, innovative products with a high level of innovation,, which can be used in medicine like personalized prosthesis for correcting orthopaedic affections, creating products based on 3D models defined, in CAD programs or byrecreating objects using reverse engineering using 3D laser scanning, with application in different domainsscans, or using 3D models created in CAD applications, the system having different applications in various fields of activity.

This subject is not approached by the scientific literature, but there are some other related subjects like using robots in milling applications [5–7] or trajectory tracking for robots in milling applications [8, 9]. This chapter is offering examples for robot engraving tasks.

2. The architecture of the system

The chapter is proposing an approach based on a software system allowing the connectivity with vertically articulated industrial robots. The system has three elements (**Figure 1**):

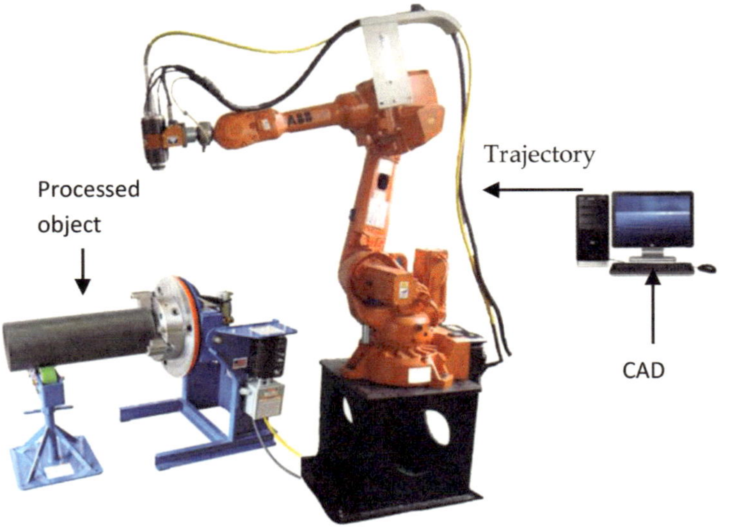

Figure 1. The architecture of the system.

1. A software application which is used to create the surfaces in 3D (CAD), allowing to:

- Import a 3D STL (STereoLitography) model of the raw object (the object before the processing).
- Import an object which describes the 3D surface. This can be done in four ways: by loading a 2D image of the trajectory of the tool; by entering a text which will be used to create an engraving on the processed object; by loading another 3D STL object which defines the robot trajectory in the space; or by loading another 3D object which represents the final product (for example after milling the raw object).
- Create an association between the initial object (without processing) and the trajectory which the robot will follow. This step can be accomplished by overlapping or mapping the object which describes the surface on the initial object: placing the text or the image over the surface of the original object if the robot task will be engraving or low depth milling, placing the 3D STL model which defines the robot trajectory in space over the initial object surface, or by placing the 3D STL model representing the final product inside the 3D model of the raw object.
- Create the trajectory which will be followed by the robot. This is done by associating the points of the surface which was described in a continuous way that can be described in Cartesian space relative to the coordinate system attached to the raw object.

2. The system used to create the robot-object reference.

This element represents a mathematical algorithm and a procedure used to create the association between the coordinate systems of the robot and respectively the coordinate system of the object. This will allow to express the 3D trajectory relative to the robot World coordinate system, the trajectory being defined in the application relative to the object coordinate system, this is done on two types of robots: ABB and OMRON (in RAPID and V_+).

3. Robot applications for trajectory tracking

This element is represented by robot applications used to control the robot to follow the computed trajectory, the applications being created to allow to specify the precision and the speed of tracking and also to specify the tool engagement angle.

3. Robot task modelling

To describe the robot task, the software application is used, the application is generating a trajectory which will be executed by the robot.

In the previous section, we seen that the application can use three different methods to obtain information about the robot task (see **Figure 2**):

- Based on a text input
- Based on a 2D image of the trajectory
- Based on a 3D STL model

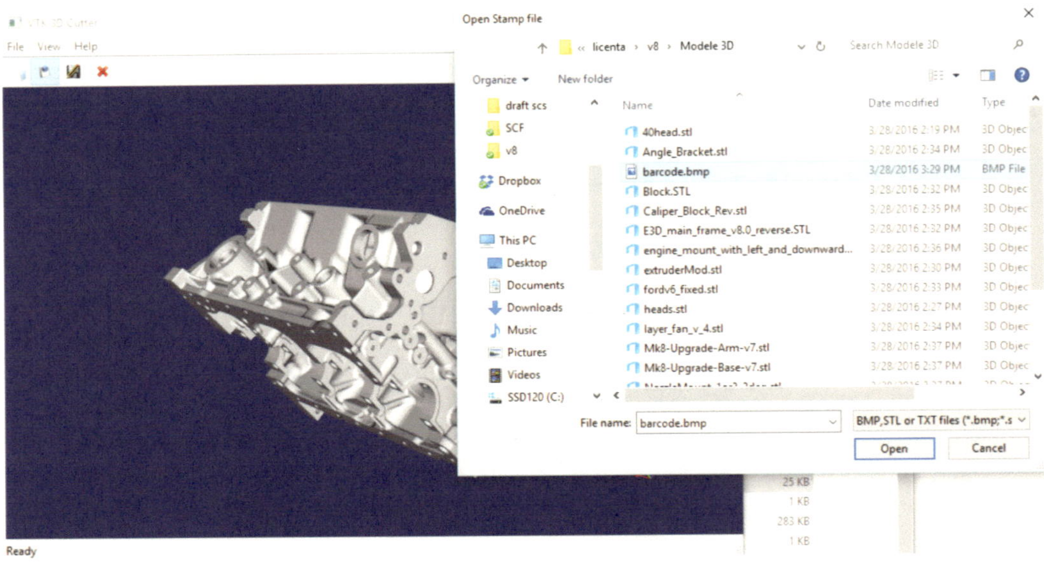

Figure 2. Loading a file in order to define the robot task.

If the user decides to use the first two methods, the input (text or image) is transformed by an image processing algorithm in order to obtain the skeleton. If the input is a text then first it is transformed into an image and then is processed. When a text is introduced into the application also the size of the characters must be introduced (in mm) and, if the task is using a milling device, then the milling depth must be also introduced. In the case of an image also the ratio mm/pixels must be introduced, and the milling depth if the object will be milled. The application is using a skeletonizing algorithm to process the input in order to obtain a simplified trajectory of the robot by thinning the imported path to the size of a line having the width of one pixel.

The skeletonizing algorithm is a process which is reducing the foreground regions in a binary image to a skeleton, which preservers the connectivity and the extent of the original object (region) while discarding most the original foreground pixels. To have a clear picture of how the algorithm is executed, we can imagine that the foreground is made of some uniform material, which is burning slow [10]. If the material is starting to burn simultaneously at all points on the boundary, the fire is advancing on the interior and at the points where is travelling from two different boundaries meets itself and is extinguishing, these points where the fire meets itself and is extinguishing are forming a so-called quench line.

The 'quench line' represents the skeleton, so in respect to this definition, we can state that the thinning operation is producing a sort of skeleton.

Another method of creating the skeleton is to compute the loci of centres of bi-tangent circles which fit entirely within the object (foreground region) which is being considered [11, 12]. **Figure 3** depicts this for a rectangular object.

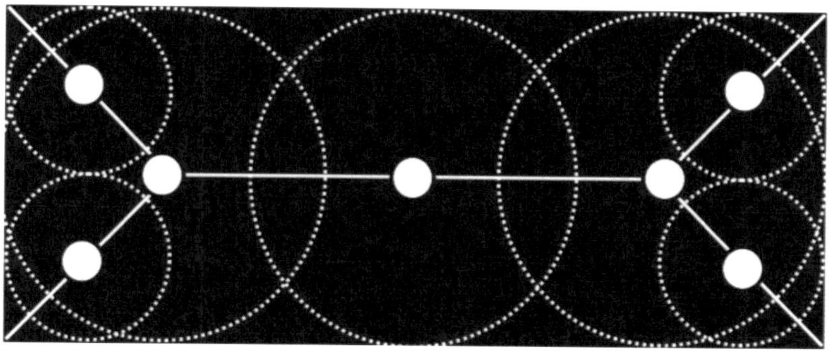

Figure 3. The skeleton created by the centres of bi-tangent circles.

In conclusion, in order to produce the skeleton, we can use two main methods. The first method is to use the thinning algorithm, which successively erodes the pixels from the frontier of object, but at the same time is preserving the end points of the line segments; until no more thinning is possible, the remnant of the object is a shape which is approximating the skeleton. Another possibility is to use an algorithm to process the image in order to obtain the distance transform. After obtaining the distance transform, the skeleton can be generated by connecting the singularities in the distance transform (i.e. discontinuities of the curvature or creases).

The majority of the actual algorithms used for skeletonization are producing skeletons with gaps; these skeletons with discontinuities cannot be used in applications where the shape must be described. This is happening because the homotopy is not preserved and important features like end points or junction points are lost. On the other hand, the thinning algorithm is generating connected skeletons, the 'quench line' being one pixel thick [13].

The thinning algorithm is making part of the class of morphological operators and is used to remove selected foreground pixels from binary images, it is similar with erosion or opening [14, 15], which are members of the same class. Thinning can be used for different applications, but is, in particular, useful for implementing the skeletonizing algorithm. For skeletonizing, thinning is used to tidy up the output of edge detectors by reducing all lines to single pixel thickness. Thinning is normally used only for processing binary images, and produces another binary image as an output. Being a morphological operator, the operation of the thinning algorithm is based on a structuring element.

Thinning is related to the hit-and-miss transform and can be expressed based on it, for example, the thinning operation applied on an image I using a structuring element S is given by:

$$thin(I, S) = I - hit\_and\_miss(I, S) \tag{1}$$

where the operator of subtraction is given by: $X - Y = X \cap NOT Y$.

The algorithm of thinning is applied by successively placing the centre of the structuring element over the pixels in the processed binary image (the structuring element is covering the entire image, from left to right and from top to bottom). For each position, the structuring element is compared with the underlying image pixels. If the structuring element matches exactly the pixels from the foreground image, then the corresponding pixel underneath the origin of the structuring element is set to background (zero), otherwise it is left unchanged. In order to have effect, the structuring element should have the origin pixel (the pixel from the centre of the structuring element) set to one or black.

Selecting the structuring element determines the conditions which should be satisfied when a foreground pixel will be set to background, and also, it determines the application for the thinning operation.

Here, we described how the image is modified after a single pass of a thinning operation; in fact, the operator is applied repeatedly until convergence (until it causes no further changes to the image). In other applications (for example, pruning), the operation is only applied for a limited number of iterations, in order to remove only few pixels.

The most common use for thinning, is to reduce the thresholded output of an edge detector like the Sobel operator, to one pixel tick lines, but in the same time preserving the length of those lines (in this way, the pixels at the ends of the lines are not affected). A simple algorithm for doing this is the following:

1. Select all pixels on the boundaries (pixels which have at least one neighbour on the background) of a foreground region (i.e. objects on a segmented image).
2. From the pixels previously selected, delete any pixel that has more than one foreground neighbour, as long as this operation is not disconnecting the region containing the pixel (i.e. split the region into two).
3. Iterate until convergence.

The procedure presented above will erode the edges of the objects in the image until no pixel can be removed (i.e. until convergence), but is not affecting the pixels which are ending the lines.

The same result can be obtained using the operator of morphological thinning which is applied until no change is produced in the image. The structuring elements which can be used are presented in **Figure 4**. The structuring elements are applied successively with all their 90° rotations (totalling $4 \times 2 = 8$ structuring elements).

This operation is executed in order to determine the octagonal skeleton of a binarized object, the skeleton being composed of the points positioned in the centres of octagons, which are tangent on the edges of the object in at least two points and fit entirely inside the object. Using this method to obtain the skeleton generates a result that is a connected skeleton having one pixel tick lines.

At each loop in the algorithm, the binary object is processed first by the structuring element from the left and then by the one from the right, and after that, using the other six 90° rotations of the two structuring elements.

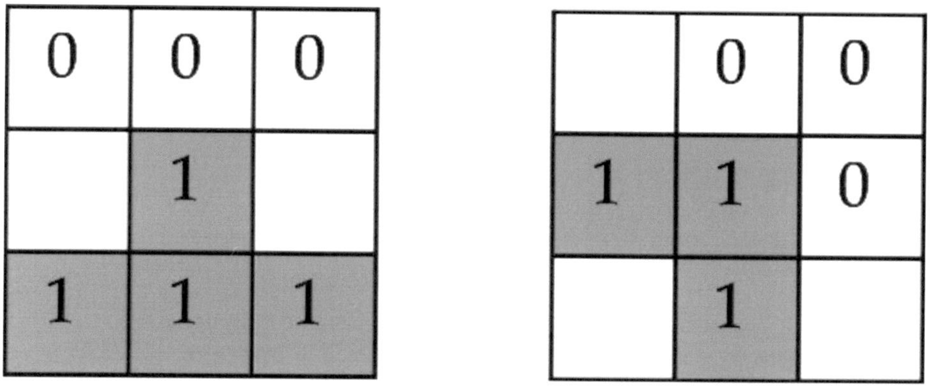

Figure 4. Structuring elements for computing the skeleton (using thinning).

The algorithm is executed in a loop until convergence [16] (until none of the eight structuring elements produces changes in the image), and the object is eroded until it remains only a set connected lines having the width of one pixel.

The thinning operation is very sensitive on perturbations, because of that the irregularities in the edges of the object will generate spurious spurs in the skeleton and this can interfere with shape recognition processes which are using the skeleton properties. In such situations, a pruning operation must be executed in order to delete spurs having a certain length or less. This method is not always efficient because small perturbations in the edge of the object can generate large spurs in the skeleton [17, 18]. Pruning is similar with thinning; the operation is based on the structuring elements presented in **Figure 5**.

In each loop, each structuring element is applied with all its 90° rotations. In contrast with thinning, the pruning algorithm is executed for only a small number of loops in order to

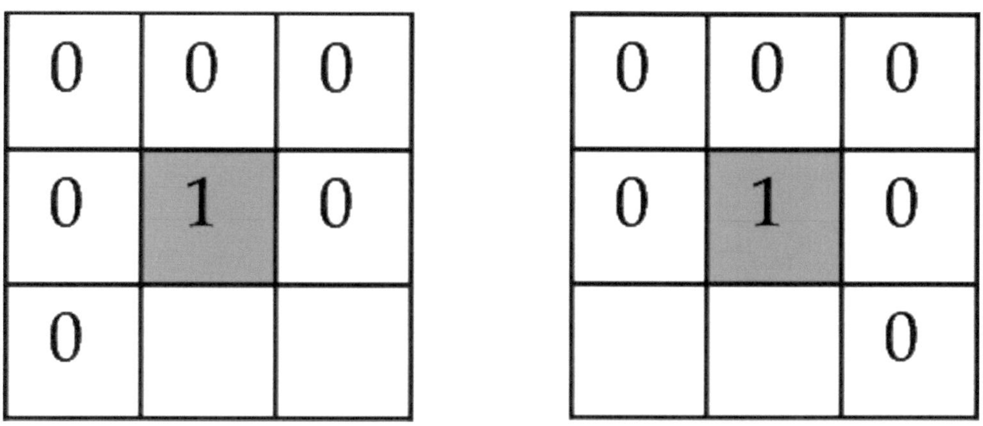

Figure 5. The structuring elements used in pruning.

remove the spurs, depending on their length. If executed until convergence, the pruning algorithm will delete all pixels except those which form closed loops.

When the image processing is done, the image which results (containing the text) can be mapped on the object which will be processed by the robot. In **Figure 6** an example is presented, the first image represents the text introduced initially, then the skeleton of the text is presented after applying the pruning algorithm, and the third image represents the same input but using a dotted font used when a continuous path is not required (for example when engraving a serial number.

The skeletonizing algorithm was also used to process input images which are describing more complex robot trajectories than text input. In **Figure 7** is presented a logo which will be engraved, the second image presents the image after has been processed by the skeletonizing and pruning algorithm, the resulted shape is composed by one pixel tick lines.

If the task is defied by using a 3D object imported into the application, the object is aligned with the initial object and placed on its surface, this is done by executing the algorithm described in **Figure 8** which is using the Iterative Closest Point (ICP) algorithm [19]. The source object represents the object which is used to define the trajectory and the destination object is the object which will be processed by the robot. **Figure 9** presents a 2D code (in the left of the figure) and in the right the code placed on the object after the execution of the ICP algorithm.

When the alignment between the two objects has been executed the ICP algorithm is applied, and then using the VTK (Visualization Toolkit) Poly Data algorithm the surface points are extracted. This algorithm which executes is filtering the pixels of based on the intersection of the two objects [20]. VTK Poly Data algorithm is implemented in python; in **Figure 10**, we can see how the algorithm is integrated with the other objects in python in order to extract the points. In **Figure 11**, we can see an example of subtraction.

Figure 6. Task definition using a text.

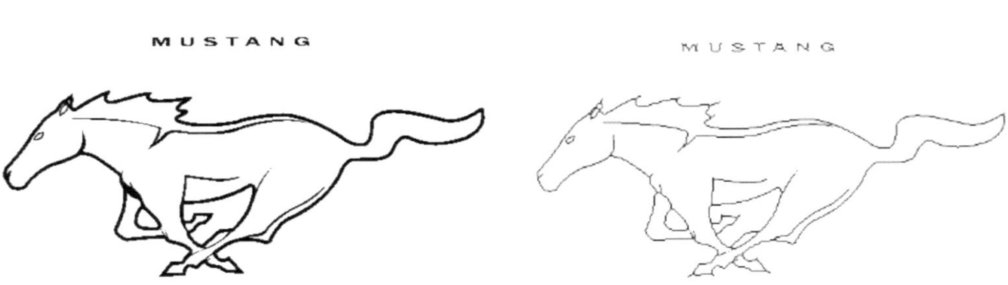

Figure 7. A logo processed for engraving.

http://dx.doi.org/10.5772/intechopen.69928

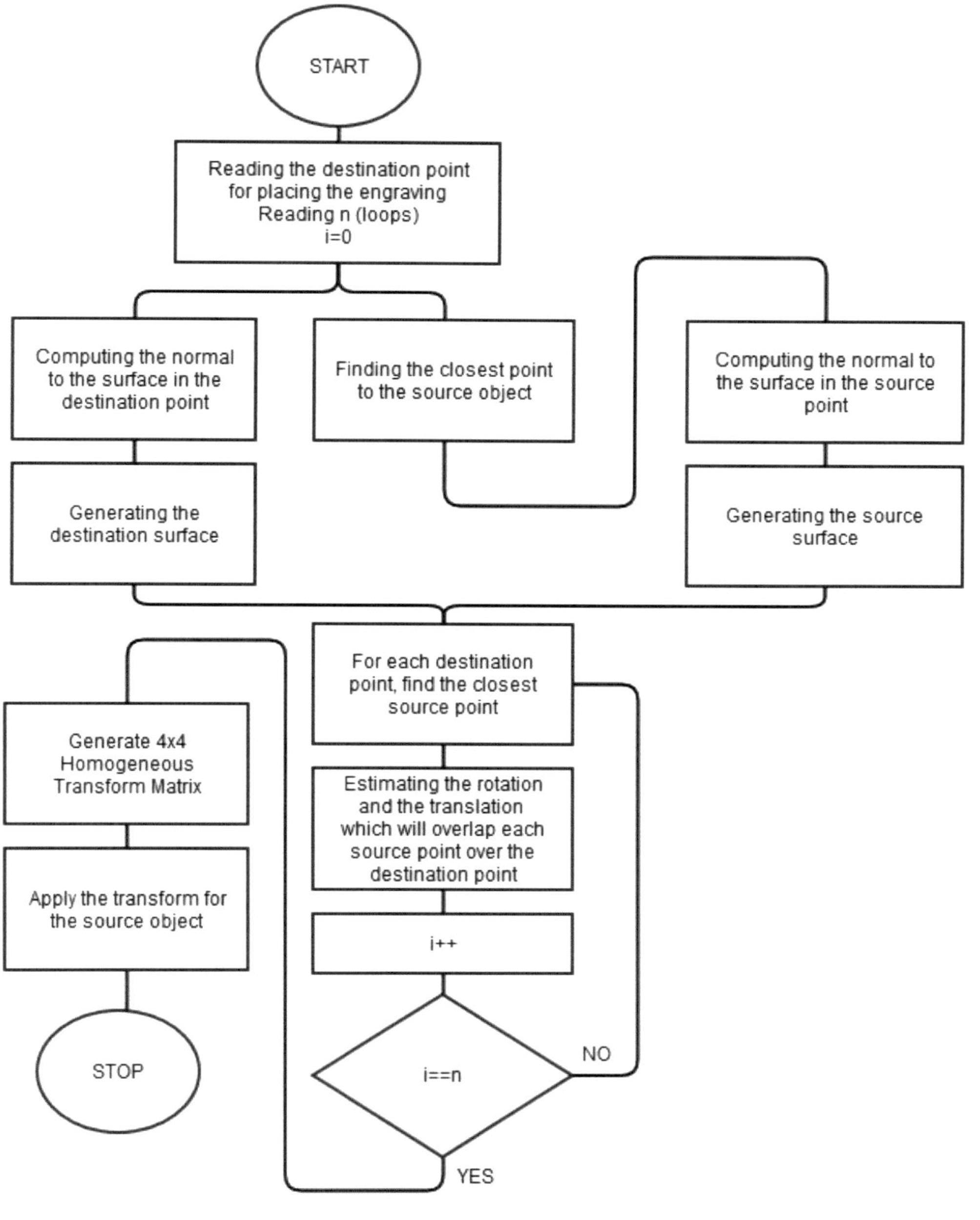

Figure 8. Algorithm for placing the engraving on the object.

The algorithm is allowing to view the result after the task has been executed (the result of the engraving), and also is extracting the points which should be accessed by the robot in order to engrave the object, this is done only when the engraving is based on points and not on continuous paths.

Figure 9. Using an STL model to define the robot task.

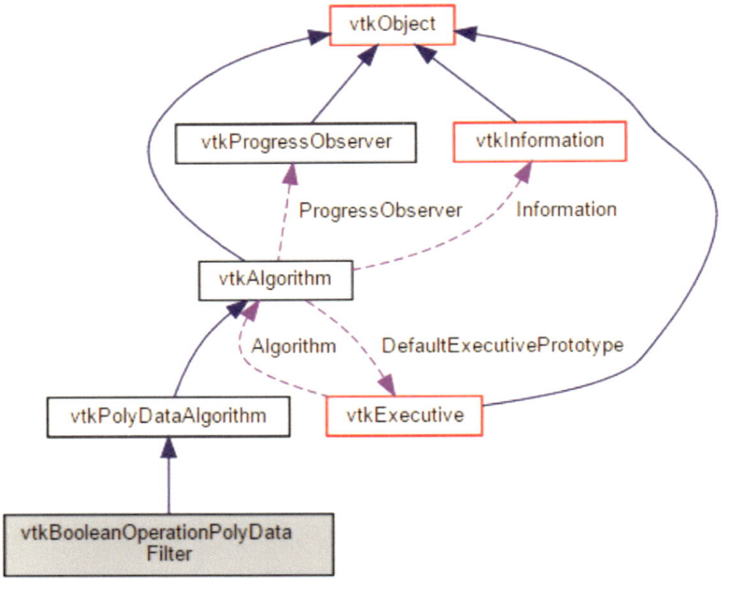

Figure 10. The integration of VTK Poly Data algorithm.

The points which are defining the trajectory are specified relative to the object coordinate system as follows:

a. In the case of using a text or an image to describe the engraving, then the skeleton is followed and the trajectory is defined by the sequence of the skeleton pixels.

b. If the case of using a dotted text, then the coordinates of the points describing the task are selected and sorted based on the distance from the coordinate system and then based on the distance from the last accessed point on the trajectory.

c. If the case of using an STL object in order to describe the trajectory, the task is described by a set of points resulted after executing the VTK Poly Data Algorithm, these points are then sorted like in case b).

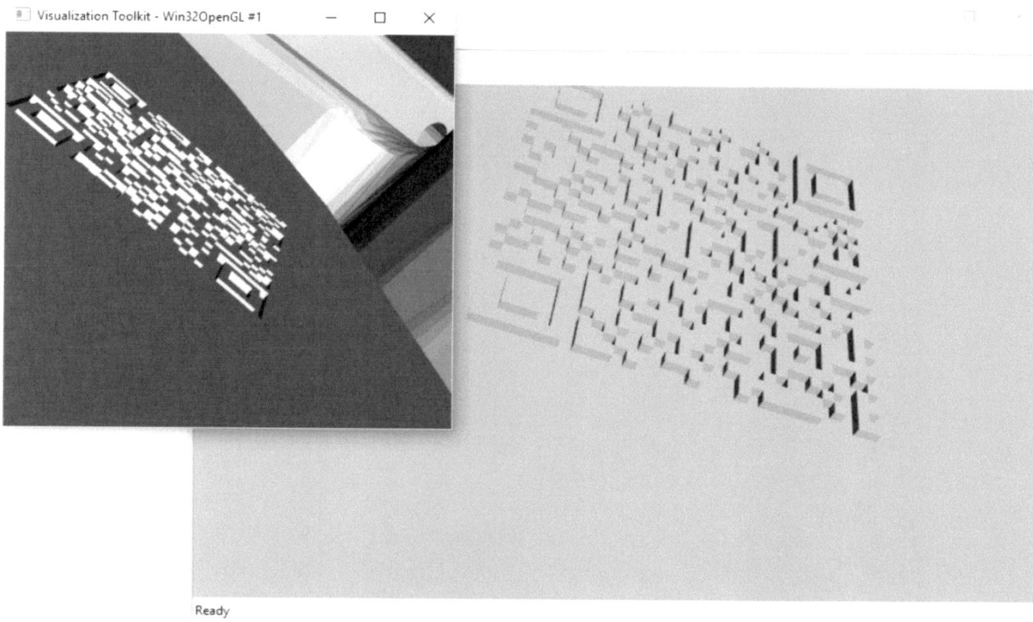

Figure 11. Subtracting a 2D code from an object surface based on VTK Poly Data algorithm.

The points which are describing the trajectory which now are placed in the order that the robot must follow are saved into a text file. This file is describing the robot task, and is sent over an ethernet connection to the robot which is using it to execute the engraving based on procedural motions.

4. Trajectory generation and tracking

In section three of the chapter we seen that the system is extracting the trajectory points which are describing the task which the robot will execute. These points are defined relative to a coordinate system which is known to the application but not to the robot system, in order that the robot to be able to execute the task, the coordinate system must be known by the robot and this is done by a robot-application calibration.

The calibration must be executed before the trajectory points are extracted and sent to the robot, and this is done by defining the object coordinate system in the application based on features which are accessible to the robot.

Figure 12 presents how the coordinate system is defined on the application. The coordinate system is defined using three points on which the robot has access:

a. The coordinate system origin is represented by point A

b. The OX axis is given by AB, where the direction is given by following the line from A to point B

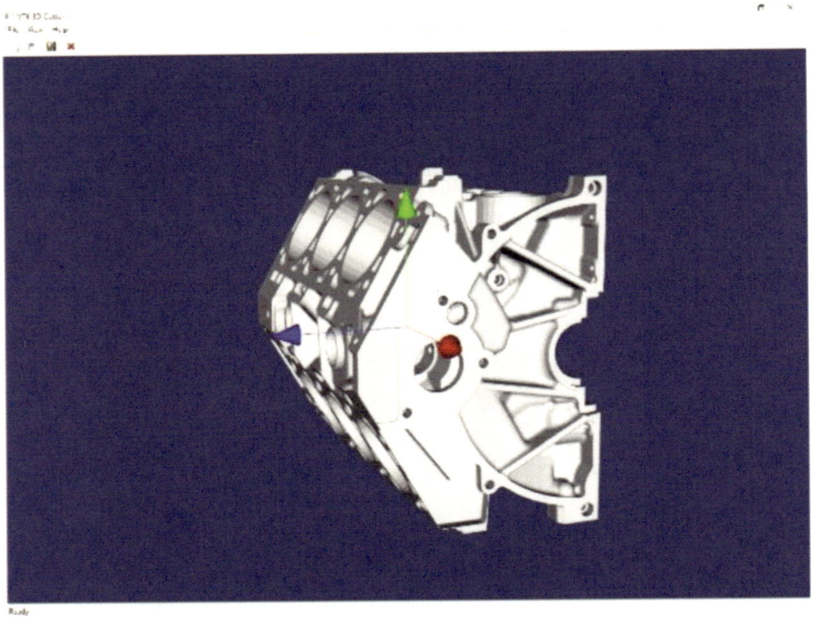

Figure 12. Placing the coordinate system on the object.

c. C is placed on the object surface, if a normal to the direction AB is generated through C, then this normal represents the OY axis. The last axis OZ, is generated by applying the right-hand rule.

Using the three points (A, B and C) trained relative to the robot World coordinate system, a frame attached to the object is defined [21] executing the command FRAME(A,B,C,A).

The resulted coordinate system is used to define the coordinates of the trajectory points in space.

The defined trajectory is followed by the robot using two methods:

a. In the case that the robot trajectory is composed by a set of points which are describing a continous path (based on a text or image) the robot enters in contact with the object using the tool and then is executing the trajectory by moving to each defined point. The tool is departed only when the next point on the trajectory is placed on a distange greater than a defined tolerance (when the trajectory is compose by multiple closed loops or discontinued lines).

b. In the case that the robot trajectory is composed by a set of points which are placed on distances greater than the defined tolerance (for example when the robot task is to engrave a dotted text, a dot code or is based on 3D STL objects), for each point on the trajectory the robot places the tool in contact with the object and then the tool is departed and is placed on the object on the next point of the trajectory.

http://dx.doi.org/10.5772/intechopen.69928

The execution time for the engraving operation is strictly related to the optimum path which will be followed by the engraving tool. A perfect trajectory will generate small holes distributed on a surface according to Ref. [22]. On the other hand, the best engraving time will be achieved if the holes are as small as possible (in terms of height).

The strategy for engraving is the drilling strategy [23]. Here, the engraving is based on drilling (in our case pointing) holes on a surface at different distances. This approach is particularly appropriated for engraving 2D codes.

In order to follow the trajectory, the set of points which are describing the 2D code must be placed on the trajectory in such way that, for example, if the tool is in the point, the next point on the trajectory will be the closest point in the code. Due to this constraint, the points on the trajectory will be arranged in ascending order based on their coordinates. A solution for finding the shortest trajectory is the algorithm Travelling Salesman Problem (TSP), this algorithm can be used for complicated engravings, in our case, for engraving a 2D code, the simplest solution is to follow each row of the 2D code.

To rearrange the points, we start from an arbitrary point (typically, this is a corner), then the closest point is added, the next point is selected from the points which have not been selected and which is the closest point relative to the last point added. The algorithm is iterated until all the points are selected.

By sorting the points based on distances does not create the trajectory, but is a useful step for determining all neighbour points for a trajectory point if we need to modify the trajectory in order to reach a point, which has been skipped when the closest method was used in order to select the points.

An example of a 2D code and the way it is engraved is given in **Figure 13**.

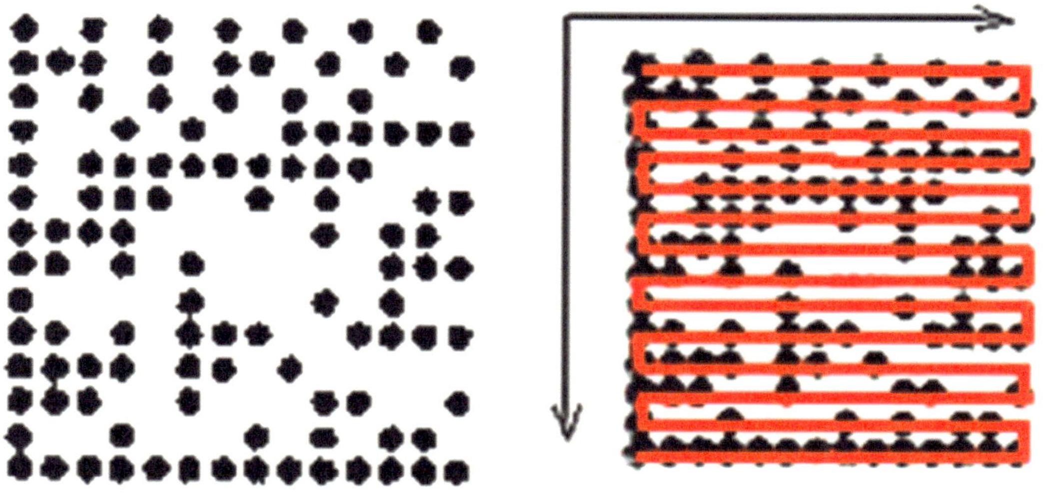

Figure 13. A 2D code and the trajectory used for engraving.

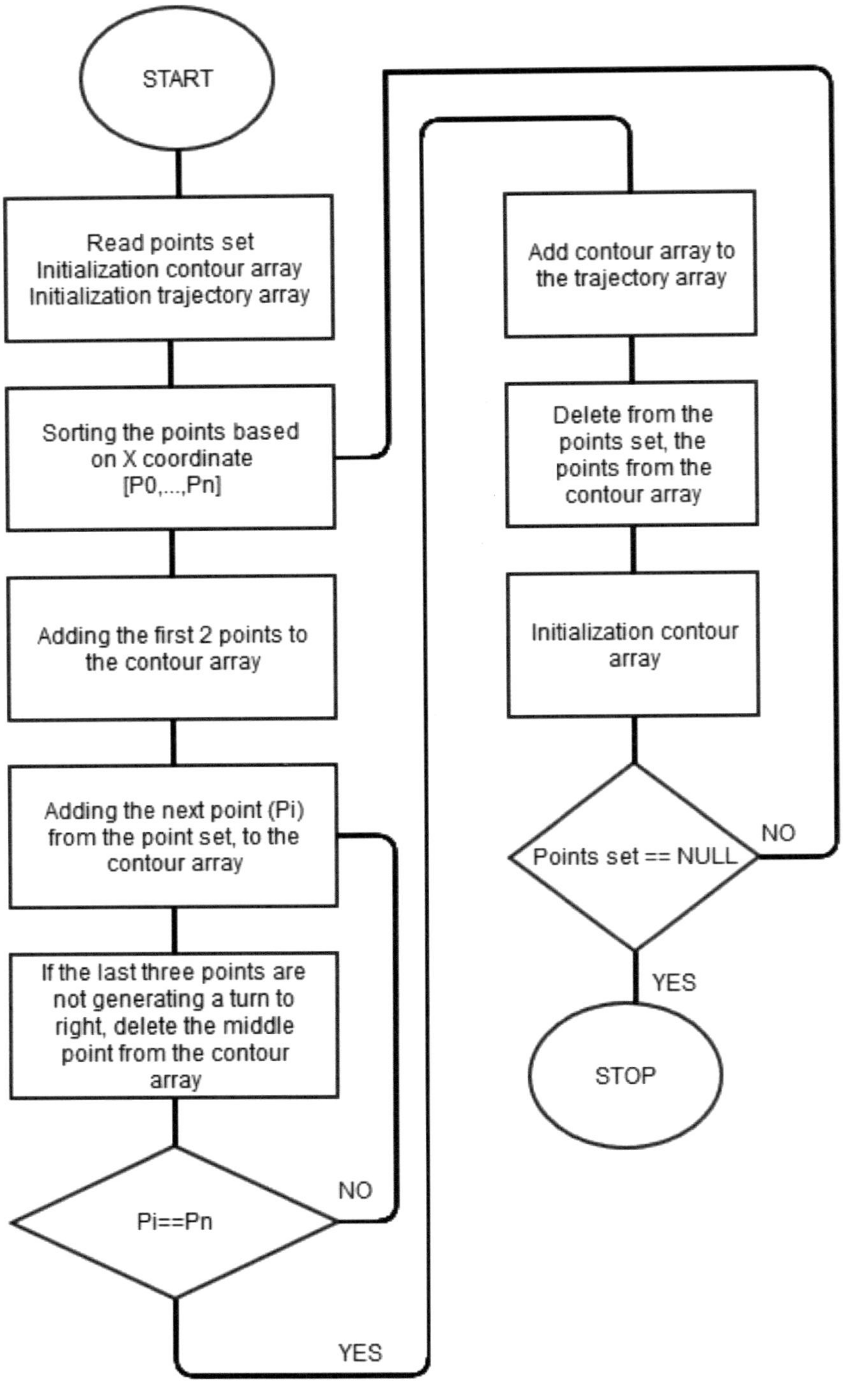

Figure 14. Trajectory generation algorithm.

http://dx.doi.org/10.5772/intechopen.69928

If the engraving is not a 2D code, we can use the same algorithm; some other algorithms which use the same principle are Graham Scan and Andrew monotone chain.

In the case of an engraving, which represents a closed surface, we can use the following algorithm [24] in order to compute the trajectory:

1. A point P_0 is selected based on its position (right, bottom).
2. The points of the engraving are sorted based of the X coordinate, obtaining a sorted array $[P_0, P_1, \ldots, P_n]$.
3. The first two points are added to the list of contour points.
4. A third point is added (in this case P_3).
5. If the last three points are not creating a turn to right, then the point in the middle (P_2) will be removed from the list. For this, we are testing if the equation$(x_2 - x_1) * (y_3 - y_1) - (y_2 - y_1) * (x_3 - x_1) < 0$ is satisfied (that means that P_2 is in the right side of P_1P_3).
6. Another point is added and the algorithm jumps to step 5 until all points are processed.

The resulted list is the exterior contour, the next step is to find the closest point to P_0 which is not part of the contour points list, this list is extracted from the engraving set of points and the algorithm is applied until no more points are left. In this way, we obtain a set of contours (which are forming the trajectory) which if they are followed by the engraving tool, the result will be the designed engraving. In **Figure 14**, the algorithm is presented.

5. Conclusion

The presented system was designed to allow the implementation of engraving services in production systems to develop new products, on demand, with a higher level of customization. The system can be considered when replacing an engraving or CNC milling machine if the task is not too complex or the execution time is not essential, so depending on the complexity of the task and the execution time the system can be a viable solution.

The system had two implementations based on two types of robots: a Viper s650 produced by Omron (former ADEPT) and an ABB IRB140 robot (see **Figure 15**). The engraving tool which was used was a Parkside hobby engraving machine equipped with a high frequency oscillating vanadium pin.

Based on the executed tests, we observed that the generated engravings are accurate if they are generated using a discrete set of points, the results are also influenced by the quality of the engraving tool which is wobbling if the amplitude of the vibrations is higher or if the trajectory is followed with a higher speed. The possible applications can be in tasks requiring short dot engravings (for example, in automotive industry).

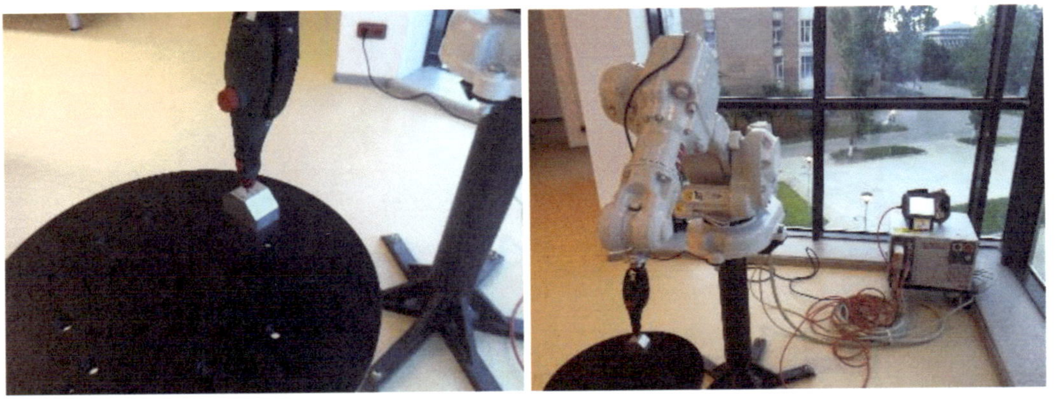

Figure 15. The implementation setup.

The integration of the system is easy and is done only by defining the robot task, adding the tool, and changing the robot programs for PC communication and task execution.

Acknowledgements

This work has been funded by University Politehnica of Bucharest, through the 'Excellence Research Grants' Program, UPB – GEX. Identifier: UPB – EXCELENȚĂ – 2016 Generarea Suprafetelor Complexe in Robotica (GSCR), 2/26.09.2016.

Author details

Silvia Anton, Florin Daniel Anton* and Mihai Constantinescu

*Address all correspondence to: anton@cimr.pub.ro

Department of Automation and Industrial Informatics, Faculty of Automatic Control and Computers, University Politehnica of Bucharest, Bucharest, Romania

References

[1] Brell-Cokcan S, Braumann J. A new parametric design tool for robot milling. In: ACADIA; ISBN: 978-1-4507-3471-4. 2010. pp. 357-363

[2] Pandremenos J, Doukas C, Stavropoulos P, Chryssolouris G. Machining with robots. In: Proceedings of 7th International Conference on Digital Enterprise Technology (DET2011); 2011; University of Patras; Laboratory for Manufacturing Systems and Automation; Athens

[3] Atmosudiro A, Keinert M, Karim A, Lechler A, Verl A, Csizar A. Productivity increase through joint space path planning for robot machining. In: UKSim-AMSS 8th European Modelling Symposium; 21-23 October 2014. Pisa, Italy: IEEE Computer Society; 2014. pp. 257-262

[4] Coelho RT, Rodella HHT, Martins VF, Barba RJ. An investigation into the use of industrial robots for machining soft and low density materials with HSM technique. ABCM. 2011;**XXIII**(3):343-350

[5] Klimchik A, Ambiehl A, Garnier S, Furet B, Pashkevich A. Efficiency evaluation of robots in machining applications using industrial performance measure. Robotics and Computer-Integrated Manufacturing. 2017;**48**:12-29

[6] Taner Tunc L, Gu Y, Burke G. Effects of minimal quantity lubrication (MQL) on surface integrity in robotic milling of austenitic stainless steel. Procedia CIRP. 2016;**45**:215-218

[7] Vosniakos GC, Matsas E. Improving feasibility of robotic milling through robot placement optimisation. Robotics and Computer-Integrated Manufacturing. 2010;**26**(5):517-525

[8] Maciej P, Grzegorz K, Konrad G, Grzegorz G, Konrad K, Janusz O. Trajectory tracking controller of the hybrid robot for milling. Mechatronics. 2016;**37**:100-111

[9] Gracia L, Garelli F, Sala A. Integrated sliding-mode algorithms in robot tracking applications. Robotics and Computer-Integrated Manufacturing. 2013;**29**(1):53-62

[10] Jain A. Fundamentals of Digital Image Processing. Prentice-Hall; USA; 1989. Chapter 9

[11] Fisher R, Perkins S, Walker A, Wolfart E. Image Processing Learning Resources [Internet]. Available from: http://homepages.inf.ed.ac.uk/rbf/HIPR2/hipr_top.htm. 2004

[12] Ballard D, Brown C. Computer Vision. Prentice-Hall; USA; 1982. Chapter 8

[13] Serra J. Image Analysis and Mathematical Morphology. Academic press; Amsterdam; Netherlands; 1992

[14] Davies E. Machine Vision: Theory, Algorithms and Practicalities. Academic Press; Amsterdam; Netherlands; 1990. pp. 149-161

[15] Haralick R, Shapiro L. Computer and Robot Vision. Vol. 1. Addison-Wesley Publishing Company; Boston; USA; 1992. pp. 168-173. Chapter 5

[16] Soille P. Morphological Image Analysis: Principle and Applications. Springer; Berlin; Germany; 1999

[17] Gonzalez R, Woods R. Digital Image Processing. Addison-Wesley Publishing Company; Boston; USA; 1992. pp. 518-548

[18] Vernon D. Machine Vision. Prentice-Hall; USA; 1991. Chapter 4

[19] Pomerleau F, Colas F, Siegwart R. A review of point cloud registration algorithms for mobile robotics. Foundations and Trends in Robotics. 2015;**4**(1):1-104

[20] Quammen C, Weigle C, Taylor II RM. Boolean operations on surfaces in VTK without external libraries. Insight Journal. 2011;**53**:1-12. DOI: http://hdl.handle.net/10380/3262

[21] Adept Technology Inc, editor. A dept V+ Reference Guide. Version 14.0, Part Number 00964-03000, Rev. B ed. San Jose, CA: Technical Publications; 2001

[22] Warkentin A, Hoskins P, Ismail F, Bedi S. Computer Aided 5-axis Machining, Computer Aided Design, Engineering and Manufacturing: System Techniques and Applications. CRC Press; London; United Kingdom; 2001. Chapter 3

[23] Mladenovic GM, Tanovic LM, Ehmann KF. Tool path generation for milling of free form surfaces with feedrate scheduling. FME Transactions. 2015;**43**(1):9-15

[24] Constantinescu M. Defining 3D robot trajectories for complex applications [dissertation]. Bucharest: UPB; 2016

Section 3

Communications in Robotics

Chapter 7

Fundamentals of Wireless Communication Link Design for Networked Robotics

Carlos Henrique Barriquello,
Flavio Eduardo Soares e Silva,
Daniel Pinheiro Bernardon, Luciane Neves Canha,
Maicon Jaderson Da Silveira Ramos and
Daniel Sperb Porto

Additional information is available at the end of the chapter

http://dx.doi.org/10.5772/intechopen.69873

Abstract

This chapter aims to present the fundamentals of the design of wireless communication links for networked robotics applications. First, we provide an overview of networked robotics applications, motivating the importance of the wireless communication link as an enabler of these applications. Next, we review the wireless communication technologies available today, discussing the existent tradeoffs between range, power, and data rate, and introducing the main concepts regarding the design of wireless communication links. Finally, we present a design example of a wireless communication link and the results obtained. We conclude the chapter with a discussion of the results and the challenges faced in the design of wireless communication links for networked robotics.

Keywords: wireless communications, networked robotics, wireless networks

1. Introduction

According to Ref. [1], service robots are defined as "reprogrammable, sensor-based mechatronic devices that perform useful service to human activities in an everyday environment." Also, service robots "perform tasks in a specific environment and should be able to perform services semi- or fully automatically." However, service robotics is different from industrial robotics, where robots are employed for the direct manufacture of goods. In service robotics, a robot performs services for humans and institutions, in an environment that often cannot be redesigned [2], and even it might be a hazardous one [3].

Examples of hazardous environments where service robots may operate include high-altitude (e.g., wall-climbing robots for inspection, painting, and cleaning of high-rise buildings [4–6]) and high-risk conditions (e.g., inspection and maintenance in nuclear and power generation industries [7–11]). Yet, in such conditions, service robots might need to be teleoperated or remotely controlled, and therefore a communication link is required [7, 12–13].

Beyond teleoperated robots (TRs), some sort of wireless communication capability is also required by autonomous robots (ARs), which are a subclass of networked robots (NRs) [14]. Accordingly, autonomous networked robots operate (possibly in group) supported by a wireless sensor network in order to fulfill their tasks [14]. Thus, the wireless sensor network extends the effective sensing range of the robots, and allows them to communicate over long distances to coordinate their activity. Yet, more recently, wireless sensor, actuator, and robot networks (WSARN) have been introduced as a means not only for extending the sensing range of the robots but also for their actuation capabilities in the surrounding environment, in order to accomplish their missions [15].

Clearly, the communication link plays an important role in networked robots applications [16], such as ubiquitous robotics [17], cloud robotics [18, 19], and remote sensing [20, 21]. For such applications, use cases may be different (e.g., robot-to-robot communication (R2R), robot-to-sensor/actuator/machine (R2S/R2A/R2M), and robot-to-cloud (R2C)) and, thus, impose different requirements for the communication (e.g., range, bit rate, latency, and energy consumption). Next, we present some of these applications.

2. Networked robotics applications

The almost ubiquitous presence of the Internet worldwide and the fast technological development in computing, sensing, and communication systems has led to envisage a new era for robotics, where robots are networked and work cooperatively with sensors, actuators, and human beings [14]. These networked robots may use external resources for computing, data gathering, sensing, learning, and working collaboratively through the "Cloud" [17, 19], or even they may "live in the Cloud" (e.g., software agents/robots) and teleoperate other robots (e.g., mobile robots and autonomous vehicles).

In **Figure 1**, we illustrate such diverse range of networked robotics applications. Networked robots may be teleoperated by a human being or by a software agent (teleoperation), sending commands to the robot(s) and receiving measurement/feedback data, thus requiring a reliable and low latency communication through the Internet. The robots also may work cooperatively, locally exchanging data in a multi-robot system and performing collaboratively a given task, thus requiring low-power and long-range wireless communication. Long-range wireless communication is also required in robotics remote-sensing systems, where mobile robots are collecting data far away in an unknown environment. On the other hand, if robots can have access to the Internet through a reliable communication link, they can possibly offload some of its processing tasks to the cloud. In cloud robotics, the robots can have access to an elastic pool of services, data, storage, and applications, extending their capabilities beyond their computing and physical constraints [19, 22].

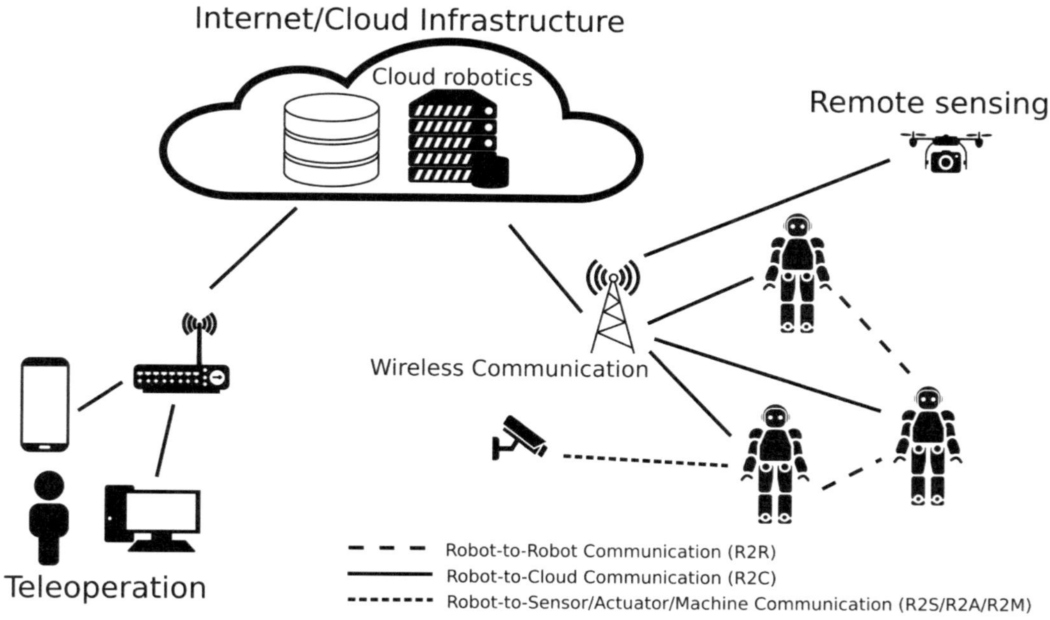

Figure 1. Representation of several envisioned networked robotics applications.

The diverse applications of networked robots impose different requirements (e.g., range, latency, reliability, bandwidth, etc.) on the communication system they rely upon, making the design of the communication network a challenging task. For example, in teleoperated robots for surgical operations, low latency and time delay are of primary concern [23], while mobile robots for outdoor mapping and the teleoperation of unmanned aerial vehicles (UAVs) require long-range communication [24, 25].

Indeed, depending on the networked robotics application, different quality-of-service (QoS) requirements (reliability, bandwidth, end-to-end delay, etc.) are imposed for the communication system design. On the other hand, there are several tradeoffs (e.g., performance, cost, range, mobility, energy consumption, etc.) involved in the design space of a communication system, which must be specifically and carefully addressed for the envisioned networked robotic application. Therefore, in the sequel, we present an overview of the design space of wireless communications systems, given its paramount importance in networked robotics [17].

3. Wireless communications overview

Wireless communication is one the major concerns in networked robotics, mainly in networked service robots [17]. However, different than wired communication, wireless is usually less reliable and more interference prone. Thus, the design of the wireless link and the wireless network is usually more challenging. The design space of a wireless communication link includes several variables, such as frequency and modulation selection, power and link-budget constraints, signal propagation characteristics, a huge set of wireless communication standards to choose from, and so on.

Wireless communication may use different portions of the electromagnetic spectrum, including radio (10 kHz to 100 MHz), microwave (100 MHz to 100 GHz), infrared (100 GHz to 400 THz), and visible light (400–790 THz). The higher the frequency, the higher the bandwidth available and more bits per second (bps) can be transmitted according to Shannon's theorem. This theorem, given in Eq. (1), relates the amount of information (C, in bps) that can be carried by a signal, such as an electromagnetic wave, with the received power (S, in Watts), the noise power (N, in Watts), and the available bandwidth (B, in Hz) for the communication system, subject to additive white Gaussian noise (AWGN)

$$C = B\log(1 + S/N) \tag{1}$$

Infrared (IR) and visible light communication (VLC) have higher bandwidth and use light-emitting diodes (LEDs) and photodetectors (PDs) for intensity modulation and direct detection (IM/DD) of the light signal [26, 27]. However, signal propagation at such high frequencies is very directional and may be easily blocked by obstacles and walls, thus requiring line of sight (LOS) for communication. Therefore, IR and VLC are more suitable for short-range indoor communication. These types of communication have been standardized by the IrDA association [28] and IEEE 802.15.7 [29, 30], respectively.

Signal propagation is better for outdoor and long-range communication in radio and microwave. However, microwave suffers much more water absorption (e.g., rain) above 6 GHz, and signal propagation is more directional. Thus, at these portions of the spectrum, bandwidth is at a premium, mainly for outdoor communication. For indoor communication, 60 GHz ISM (industrial, scientific, and medical) unlicensed band is worldwide available and its use for Wi-Fi communication has been standardized as the IEEE 802.11ad specification [31, 32], offering up to 7 Gbps throughput and allowing the use of very small and low-cost antennas [33, 34].

For short-range communication, IEEE 802.11 (Wi-Fi) is an interesting option for applications requiring high throughput (e.g., video streaming). It may operate also in the worldwide available ISM bands in 2.4 (IEEE 802.11b/g/n) and 5.8 GHz (IEEE 802.11ac). However, IEEE 802.11 is more power and energy hungry than other wireless standards such as Bluetooth Low Energy (BTLE) [35] and IEEE 802.15.4 [36] standards, which are targeted for low-power and low-rate applications.

IEEE 802.15.4 is a standard for low-rate (e.g., 250 kbps in 2.4-GHz band) and low-power (<100 mW) wireless communication. It is available worldwide in 2.4 GHz and several other bands with availability depending on the country/region of operation [36]. Also, it has been adopted by other application and networking standards, such as ZigBee [37] and 6LoWPAN [38]. On the other hand, the BLTE standard has similar performance characteristics of IEEE 802.15.4, such as low power (<100 mW) and low rate (<1 Mbps), but operates solely in 2.4-GHz band. It is promoted by Bluetooth Special Interest Group and is widely used in smartphones and tablets.

However, both IEEE 802.15.4 and BLTE are designed for short-range communication (<1 km), and thus they are not targeted for applications requiring long-range communications.

Long range is usually required in outdoor wireless communications. In such cases, lower frequencies are preferable due to lower path loss (PL), which is the attenuation of signal power between the transmitter (Tx) and the receiver (Rx), measured in decibels (dB), in a given wireless communication link. Free space path loss (FSPL), which does not account for obstacles and reflections, and depends solely on frequency (f in GHz) and distance (d in km), is given by Eq. (2) in dB

$$Lp = 92.45 + 20\log_{10}(d) + 20\log_{10}(f) \tag{2}$$

At a long distance, a more realistic model is the two-ray ground reflection model, which accounts also for the reflection of the signal on the ground. In this model, both the reflected wave component and the direct LOS wave component are considered. In this case, the path loss also depends on the gain (G) and the heights (ht, hr) of the transmitter and receiver antennas, and is given by Eq. (3) that is only valid for large d (i.e., $d \gg \sqrt{hr \times ht}$).

$$Lp = 40\ \log_{10}(d) - 10\log_{10}(Ght^2hr^2) \tag{3}$$

In fact, the heights of the antennas not only influence the path loss but also limit the maximum line-of-sight (LOS) distance, which is dependent on the earth's curvature. Accordingly, the maximum LOS distance, in meters, is given by Eq. (4), where R is earth's radius and equals 6365 km. For instance, considering antennas placed at 1-m height above the ground, the maximum LOS is 7136 m

$$LOS\text{max} = \sqrt{2 \times ht \times R} + \sqrt{2 \times hr \times R} \tag{4}$$

In practice, however, even if the distance is lower than the maximum LOS, communication will not be possible if the received signal power is too low such that it cannot be distinguished from the noise present at the receiver. This relation is known as signal-to-noise ratio (SNR). Therefore, for a given modulation and communication bit rate R, there is a minimum SNR required by the receiver in order to achieve a desired bit error rate (BER) or packet error rate (PER). Such a minimum required SNR is usually defined in terms of the receiver sensitivity, which is the absolute input power level required to not exceed 1% BER or PER at the receiver. Thus, the receiver sensitivity, which is specified in the receiver's datasheet, is the input power level that gives the required minimum SNR for the wireless communication link.

The design of a reliable communication must be done in order to assure that the received power (*Rx power*) is at least equal to the receiver sensitivity (*Rx sens*) plus some safety margin (known as *link margin*), according to Eq. (5). The link margin must be chosen in order to have enough received power even in case of signal attenuation due to mobility and multipath propagation, which is known as fading and is the result of the destructive interference among the waves that travel through different paths and reach the receiver with different delays. As a rule of thumb, link margin, also known as fade margin, is usually set to 10–30 dB, depending on the desired link reliability

$$Rx\ power\ (dBm) = Rx\ sens\ (dBm) + Link\ margin\ (dB) \tag{5}$$

The received power can be determined taking into account the transmitted power (*Tx power*) plus the gains and losses through the wireless link, according to Eq. (6). Then, combining Eqs. (5) and (6), putting in the gains (*Gr, Gt*) of the receiver and the transmitter antennas, the path loss (in dB), and rearranging the terms, one obtains Eq. (7), which is the "link power budget" or "link-budget" equation. Therefore, the link budget, given in Eq. (8), is the maximum allowed amount of power that can be lost through the wireless link due to path loss and fading, and still maintaining the communication link working. Equation (8) can also be written as Eq. (9), where the term EIRP (dBm) is known as the *effective isotropic radiated power*

$$Rx\ power\ (dBm) = Tx\ power\ (dBm) + gains\ (dB) - losses\ (dB) \tag{6}$$

$$Path\ loss(dB) + Link\ margin(dB) = Tx\ power(dBm) + Gt(dB) + Gr(dB) - Rx\ sens(dBm) \tag{7}$$

$$Link\ budget\ (dB) = Tx\ power\ (dBm) + Gt\ (dB) + Gr\ (dB) - Rx\ sens\ (dBm) \tag{8}$$

$$Link\ budget\ (dB) = EIRP\ (dBm) + Gr\ (dB) - Rx\ sens\ (dBm) \tag{9}$$

For wireless systems operating in unlicensed ISM bands, the EIRP is limited by local regulatory agencies, such as the Federal Communications Commission (FCC) in the United States, the European Telecommunications Standards Institute (ETSI) in Europe, the Association of Radio Industries and Business (ARIB) in Japan, and the National Telecommunications Agency (ANATEL) in Brazil.

Note that the limitation imposed to the EIRP is very important in the design of a long-range communication link. If the link is bidirectional and the same antenna is used for transmission and reception, the maximum achievable link budget will be basically determined by the receiver sensitivity, as can be noticed from Eq. (9). Therefore, the selection of the receiver with enough sensitivity is a crucial step in the design of a long-range wireless communication system.

Basically, there are two main approaches in order to achieve long-range communication: ultra narrowband (UNB) radio-frequency (RF) and wideband spread spectrum (SS). Both approaches aim to increase the receiver sensitivity at the cost of reducing the effective data rate. Ultra narrowband RF is a technique for wireless communication where the bandwidth used is very small compared to carrier frequency (i.e., $\frac{\Delta f}{f} \ll 1$). Therefore, the signal energy is concentrated in this narrow band and the thermal noise is reduced, ultimately improving the SNR.

The reduction of thermal noise due to the reduction of the bandwidth can be noted from Eq. (10), which shows that thermal noise (N) is proportional to the bandwith B and the temperature T. In Eq. (10), the proportionality constant k is Boltzmann's constant and equals 1.38×10^{-23} J/K. Therefore, in room temperature (T = 290 K), the noise power in dBm can be calculated by Eq. (11). For example, in a communication system operating with a bandwidth of 10 kHz, the noise floor is -134 dBm

$$N = k \times T \times B \tag{10}$$

$$N(dBm) = -174 + 10\log_{10}(B) \tag{11}$$

The improvement of the SNR due to UNB results in the improvement of the receiver sensitivity, and thus allows for higher link budgets and longer ranges. However, the throughput is also reduced, making UNB suitable only for low-power and low-rate networks, such as low-power wide area networks (LPWAN) [39] and low throughput networks (LTN) [40]. For instance, Sigfox [41], which is a LWPAN/LTN provider operating in unlicensed bands (915 MHz in US and 868 MHz in Europe), employs UNB communication with a very small bandwidth of 100 Hz and a bit rate of 100 bps. With such a low bandwidth, the spectrum is efficiently used and the noise power is very low (around −150 dBm at 290 K), allowing a receiver to demodulate an extremely low-power signal of −142 dBm [42]. In [39], it is reported that a UNB test link of 25 km was deployed successfully with a transmission power of 14 dBm and an SNR exceeding 20 dB.

Another alternative for increasing the receiver sensitivity is through the use of some spread spectrum technique. In such technique, the bandwidth used to transmit the signal is β times larger than the minimum required, where β is known as the spreading factor (SF), repetition factor, or processing gain. The effect of spreading factor is the reduction of the spectral efficiency η, which is the ratio between the bit rate R and the bandwidth B, while maintaining the bandwidth B constant [43]. Ultimately, this reduces the minimum required SNR, thus lowering the receiver sensitivity, according to Eq. (12), where $SNR_{\min}$ (dB) is the SNR required without the spreading technique and SNR_{β} (dB) is the SNR required with the use of the spreading factor β [44]

$$SNR_{\beta}(dB) = SNR_{\min}(dB) - 10\log_{10}(\beta) \tag{12}$$

The spreading technique is used by a state-of-the-art modulation known as LoRa, which is promoted by the LoRa Alliance for LPWAN deployments based on the LoRaWAN specification [45, 46]. LoRa is based on chirp spread spectrum (CSS) modulation, which uses wideband linear frequency-modulated pulses whose frequency varies linearly over time in order to encode information [47, 48]. LoRaWAN has been designed to operate in several license-exempt bands, including 868-MHz band in Europe and 915-MHz band in US, with configurable bandwidth between 125, 250, and 500 kHz and configurable spreading factor between 7 and 12. According to the chosen spreading factor, the data rates range from 336 to 48 kbps [49], as calculated by Eq. (13), where B is the bandwidth and R is the data rate

$$R = \frac{B}{2^{SF}} \times SF \tag{13}$$

LoRa can offer a receiver sensitivity lower than −130 dBm, thus allowing long-range communication links [47]. In Ref. [46], working LoRa links operating in 868-MHz band with ranges up to 15 km over the ground and 30 km over water were reported, when using the highest spreading factor (SF = 12) and the maximum allowed transmission power (14 dBm), with the bandwidth set to 125 kHz. However, in order to achieve those long ranges, the bit rate was reduced to only 293 bps, thus showing a clear tradeoff between the distance and the data rate in the design of a wireless communication link. **Table 1** shows the achievable data rates, expected range, and time on air for LoRa according to the spreading factor.

Spreading factor (at 125 kHz)	Bit rate (bps)	Expected range (km)	Time on air (ms) (for 10 bytes payload)
SF7	5470	2	56
SF8	3125	4	100
SF9	1760	6	200
SF10	980	8	370
SF11	440	11	700
SF12	290	14	1400

Bandwidth set to 125 kHz with coding rate 4/5 and 1% PER.

Table 1. Data rates, time on air, and expected range (depending on propagation conditions) for LoRa according to the spreading factor.

4. Design example of a wireless communication link

In this study, a scenario is considered where long-range communication is required. Thus, it is a representative case for networked robotics where large distances are involved, such as in the monitoring of inspection robots for power transmission lines [13] and pipes [50], UAVs control, remote sensing, and navigation of mobile robots in large open fields. The choice of such scenario (i.e., a large outdoor area) was motivated by its importance for some envisioned applications of networked robots (e.g., smart cities, smart grids, agriculture, mining and military applications, etc.) and by the challenge it represents for the design of the wireless communication link (e.g., link budget, range, fading, obstacles, mobility, etc.).

The selected frequency band is 915-MHz band, which is an ISM band in Americas with a maximum EIRP of 36 dBm for spread spectrum systems [51]. Therefore, path losses at 915 MHz for 130 different positions, with ranges varying from 1 to 40 km, have been obtained through simulations with the software LINKPlanner [50], as shown in **Figure 2**. LINKPlanner performs the calculations from the International Telecommunication Union (ITU) recommendations ITU-R P.526-10 and ITU-R P.530-12 to predict NLOS (non-line-of-sight) and LOS paths for anywhere in the world [52]. As can be noticed from **Figure 4**, path losses vary from 90 up to 160 dB, thus requiring high link budgets in order to cover all possible ranges in the scenario of study.

Given the high link budgets required and the limitation of EIRP, the receiver sensitivity must be as low as possible, such as the availability in UNB and LoRa transceivers. Therefore, in this design example, the SX1272 LoRa transceiver has been selected, which can transmit at up to +20-dBm output power and has a sensitivity as low as −137 dBm at 125 kHz with a spreading factor of 12 [53].

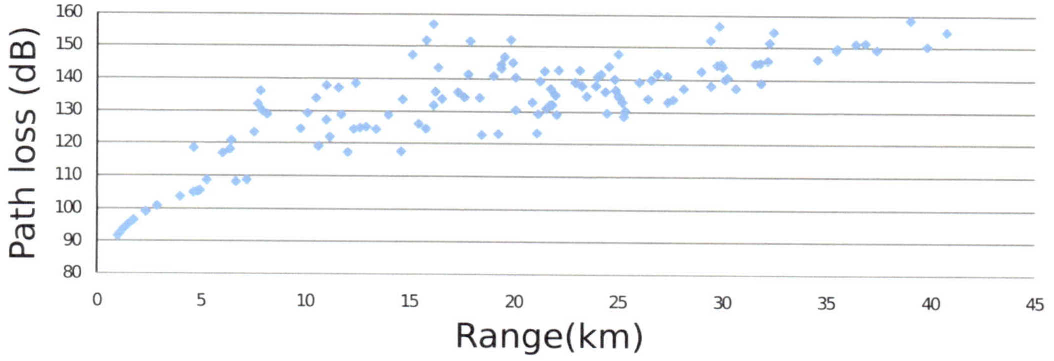

Figure 2. Path losses (dB) versus range (km) at 915 MHz obtained through simulations with LINKPlanner [50]. Antenna heights ht = 50 m and hr = 1 m.

Using the maximum transmit power with unitary gain antennas, the link budget is limited to 157 dB, thus 3 dB below the minimum required to cover all possible ranges in the scenario of study (i.e., 160 dB). However, with 6-dBi gain antenna at the transmitter and 9-dBi gain antenna at the receiver, 15-dB gain is added to the system, leaving still 12 dB of link margin and covering all considered ranges, as shown in **Figures 3** and **4**. In **Figure 5**, the distribution of the required spreading factors with and without a 10-dB signal fading is shown.

Yet, as the maximum EIRP is 36 dBm with +30 dBm of transmit power and 6 dBi of antenna gain, there is still a margin to add more 10 dB of link budget. This can be done using another transceiver with more output power capability (i.e., up to +30 dBm) or, preferably in case of a bidirectional link, through the use of antennas with higher gains. The reason is that the antenna not only increases the EIRP but also adds gain for signal receiving, however, at the cost of less area coverage due to the increase of the antenna directivity.

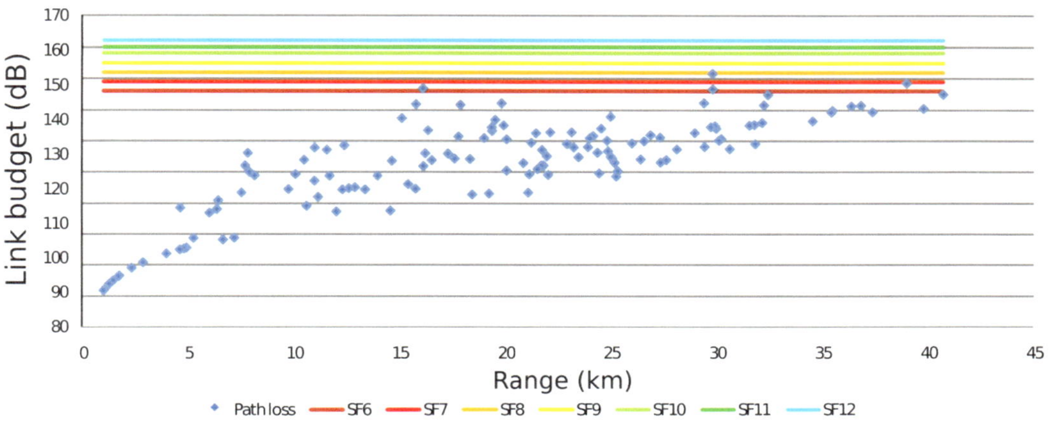

Figure 3. Link budgets (dB) versus range (km) with LoRa transceiver at 915 MHz for different spreading factors (6–12). Antenna heights ht = 50 m and hr = 1 m, with gains Gt = 6 dBi and Gr = 9 dBi.

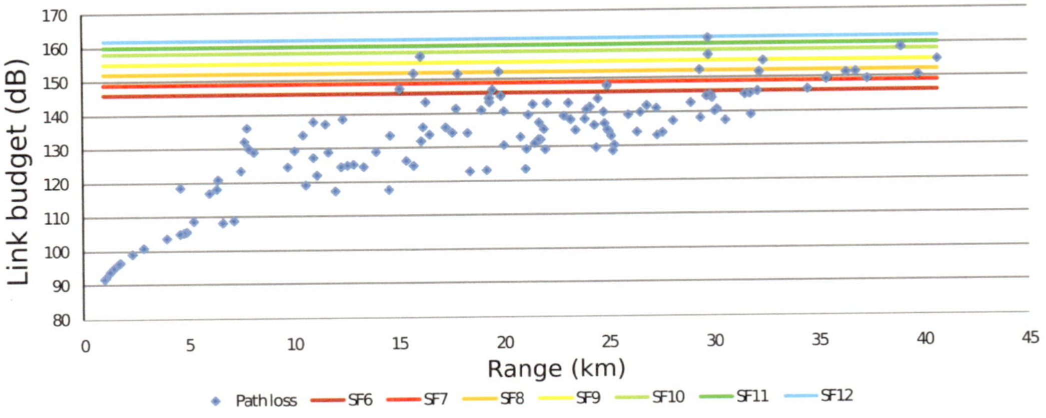

Figure 4. Link budgets (dB) versus range (km) with LoRa transceiver at 915 MHz for different spreading factors (6–12), with 10-dB link margin for signal fading. Antenna heights ht = 50 m and hr = 1 m, with gains Gt = 6 dBi and Gr = 9 dBi.

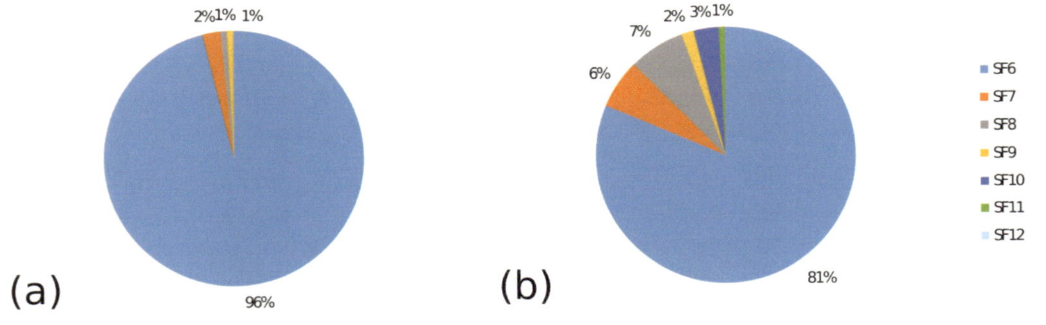

Figure 5. Distribution of spreading factors without (a) and with (b) 10-dB signal fading. LoRa transceiver at 915 MHz with 125-kHz bandwidth. Antenna heights ht = 50 m and hr = 1 m, with gains Gt = 6 dBi and Gr = 9 dBi.

In **Figure 6**, the time on air for different spreading factors and payload sizes ranging from 10 to 50 bytes is shown. It can be noticed that the time on air is longer for higher spreading factors (i.e., lower bit rates) and longer payload sizes. Thus, it is important to use, whenever possible, small packets and higher bit rates for applications that require low communication delay, such as robot teleoperation.

5. Conclusion

In this chapter, we have presented the fundamentals of the design of wireless communication links. We have discussed the importance of this topic for networked robotics applications,

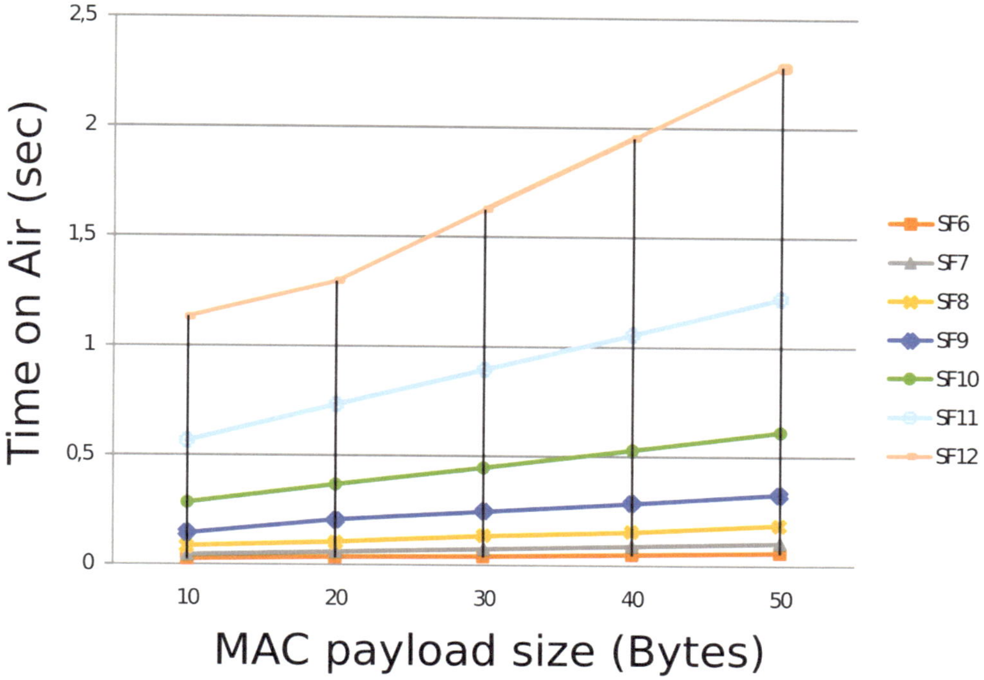

Figure 6. Time on air with different spreading factors and MAC payload sizes. LoRa transceiver at 915 MHz with 125-kHz bandwidth and coding rate 4/5.

which usually require some sort of wireless communication. After reviewing the fundamentals of wireless communication and several wireless communication technologies, we have shown a design example of a wireless communication link based on Lora modulation, one of the latest available wireless technologies for long-range communication. The obtained results allow us to show the existent tradeoff between communication range, data rate, and delay in the design of a wireless communication link. Although long-range communication is possible, the data rate needs to be reduced, negatively affecting the communication delay. Therefore, networked robotics applications that require long-range communications must be able to work properly with low data rates (e.g., using data compression algorithms) and withstand temporarily network disconnection due to signal fading (e.g., using opportunistic communication [54]).

Acknowledgements

The authors would like to thank the technical and financial support of RGE Sul Power Utility by project "Solução Inovadora para Gerenciamento Ativo de Sistemas de Distribuição" (P&D/ANEEL), Coordination for the Improvement of High Level Personnel (CAPES) and the National Center of Scientific and Technological Development (CNPq).

Author details

Carlos Henrique Barriquello[1]*, Flavio Eduardo Soares e Silva[2], Daniel Pinheiro Bernardon[1], Luciane Neves Canha[1], Maicon Jaderson Da Silveira Ramos[2] and Daniel Sperb Porto[2]

*Address all correspondence to: barriquello@gmail.com

1 Federal University of Santa Maria, Santa Maria, Brazil

2 RGE Sul, Brazil

References

[1] Kawamura K, Pack RT, Bishay M, Iskarous M. Design philosophy for service robots. Robotics and Autonomous Systems. 1996;**18**(1):109-116, ISSN 0921-8890. DOI: http://dx.doi.org/10.1016/0921-8890(96)00005-X

[2] Decker M, Fischer M, Ott I. Service robotics and human labor: A first technology assessment of substitution and cooperation. Robotics and Autonomous Systems. January 2017;**87**:348-354, ISSN 0921-8890. DOI: http://dx.doi.org/10.1016/j.robot.2016.09.017

[3] Luk BL, Cooke DS, Galt S, Collie AA, Chen S. Intelligent legged climbing service robot for remote maintenance applications in hazardous environments. Robotics and Autonomous Systems. 30 November 2005;**53**(2):142-152. ISSN 0921-8890. DOI: http://dx.doi.org/10.1016/j.robot.2005.06.004

[4] Nansai S, Mohan RE. A survey of wall climbing robots: Recent advances and challenges. Robotics. 2016;**5**:14

[5] Guan Y, et al. A modular biped wall-climbing robot with high mobility and manipulating function. IEEE/ASME Transactions on Mechatronics. December 2013;**18**(6):1787-1798. DOI: 10.1109/TMECH.2012.2213303

[6] Schmidt D, Berns K. Climbing robots for maintenance and inspections of vertical structures—A survey of design aspects and technologies. Robotics and Autonomous Systems. December 2013;**61**(12):1288-1305. ISSN 0921-8890. DOI: http://dx.doi.org/10.1016/j.robot.2013.09.002

[7] Luk BL, Liu KP, Collie AA, Cooke DS, Chen S. Tele-operated climbing and mobile service robots for remote inspection and maintenance in nuclear industry. Industrial Robot: An International Journal. 2006;**33**(3):194-204

[8] Bogue R. Robots in the nuclear industry: A review of technologies and applications. Industrial Robot: An International Journal. 2011;**38**(2):113-118

[9] Li Z, Ruan Y. Autonomous inspection robot for power transmission lines maintenance while operating on the overhead ground wires. International Journal of Advanced Robotic Systems. 2010;**7**(4):111-116

[10] Pagnano A, Höpf M, Teti R. A roadmap for automated power line inspection. Maintenance and Repair, Procedia CIRP. 2013;**12**:234-239. ISSN 2212-8271. DOI: http://dx.doi.org/10.1016/j.procir.2013.09.041

[11] Rogério Sales G, Carvalho JCM. Review and latest trends in mobile robots used on power transmission lines. International Journal of Advanced Robotic Systems (Print). 2013;**10**:1-14

[12] Schilling K, Driewer F. Remote control of mobile robots for emergencies. IFAC Proceedings Volumes. 2005;**38**(1):65-70. ISSN 1474-6670. DOI: http://dx.doi.org/10.3182/20050703-6-CZ-1902.01166

[13] van Osch M, Bera D, van Hee K, Koks Y, Zeegers H. Tele-operated service robots: ROSE. Automation in Construction. 1 April 2014;**39**:152-160. ISSN 0926-5805. DOI: http://dx.doi.org/10.1016/j.autcon.2013.06.009

[14] Sanfeliu A, Hagita N, Saffiotti A. Network robot systems. Robotics and Autonomous Systems. 31 October 2008;**56**(10):793-797. ISSN 0921-8890. DOI: http://dx.doi.org/10.1016/j.robot.2008.06.007

[15] Curiac D-I. Towards wireless sensor, actuator and robot networks: Conceptual framework, challenges and perspectives. Journal of Network and Computer Applications. March 2016;**63**:14-23. ISSN 1084-8045. DOI: http://dx.doi.org/10.1016/j.jnca.2016.01.013

[16] Isler V, Sadler B, Preuchil L, Nishio S. Networked robots [TC spotlight]. IEEE Robotics & Automation Magazine. September 2015;**22**(3):25-29. DOI: 10.1109/MRA.2015.2452172

[17] Chibani A, Amirat Y, Mohammed S, Matson E, Hagita N, Barreto M. Ubiquitous robotics: Recent challenges and future trends. Robotics and Autonomous Systems. November 2013;**61**(11):1162-1172. ISSN 0921-8890. DOI: http://dx.doi.org/10.1016/j.robot.2013.04.003

[18] Qureshi B, Koubâa A. Five traits of performance enhancement using cloud robotics: A survey. Procedia Computer Science. 2014;**37**:220-227. ISSN 1877-0509. DOI: http://dx.doi.org/10.1016/j.procs.2014.08.033

[19] Kehoe B, Patil S, Abbeel P, Goldberg K. A survey of research on cloud robotics and automation. IEEE Transactions on Automation Science and Engineering. April 2015;**12**(2):398-409. DOI: 10.1109/TASE.2014.2376492

[20] Rathje EM, Franke K. Remote sensing for geotechnical earthquake reconnaissance. Soil Dynamics and Earthquake Engineering. December 2016;**91**:304-316. ISSN 0267-7261. DOI: http://dx.doi.org/10.1016/j.soildyn.2016.09.016

[21] Toth C, Józków G. Remote sensing platforms and sensors: A survey. ISPRS Journal of Photogrammetry and Remote Sensing. May 2016;**115**:22-36. ISSN 0924-2716. DOI: http://dx.doi.org/10.1016/j.isprsjprs.2015.10.004

[22] G. Hu, W. P. Tay and Y. Wen, "Cloud robotics: architecture, challenges and applications," in IEEE Network, vol. 26, no. 3, pp. 21-28, May-June 2012. DOI: 10.1109/MNET.2012.6201212

[23] M. J. H. Lum et al., "Teleoperation in surgical robotics – network latency effects on surgical performance," 2009 Annual International Conference of the IEEE Engineering in Medicine and Biology Society, Minneapolis, MN, 2009, pp. 6860-6863. DOI: 10.1109/IEMBS.2009.5333120

[24] Gebre BA, Men H, Pochiraju K. Remotely operated and autonomous mapping system (ROAMS). In: 2009 IEEE International Conference on Technologies for Practical Robot Applications; Woburn, MA; 2009. pp. 173-178. DOI: 10.1109/TEPRA.2009.5339624

[25] Mammadov E, Gueaieb W. Long-range communication framework for multi-agent autonomous UAVs. In: 2014 International Conference on Unmanned Aircraft Systems (ICUAS); Orlando, FL; 2014. pp. 395-403. DOI: 10.1109/ICUAS.2014.6842279

[26] Kahn JM, Barry JR. Wireless infrared communications. Proceedings of the IEEE. February 1997;**85**(2):265-298. DOI: 10.1109/5.554222

[27] Mossaad MSA, Hranilovic S, Lampe L. Visible light communications using OFDM and multiple LEDs. IEEE Transactions on Communications. November 2015;**63**(11)4304-4313. DOI: 10.1109/TCOMM.2015.2469285

[28] Irda Association [Internet]. 2011. Available from: irda.org. [Accessed: 2017–01–18]

[29] IEEE Standard for Local and Metropolitan Area Networks—Part 15.7: Short-range wireless optical communication using visible light. IEEE Std 802.15.7-2011. September 6, 2011;**1**(1):1-309. DOI: 10.1109/IEEESTD.2011.6016195. Available from: standards.ieee.org/getieee802

[30] Rajagopal S, Roberts RD, Lim SK. IEEE 802.15.7 visible light communication: Modulation schemes and dimming support. IEEE Communications Magazine. March 2012;**50**(3):72-82. DOI: 10.1109/MCOM.2012.6163585

[31] Nitsche T, Cordeiro C, Flores AB, Knightly EW, Perahia E, Widmer JC. IEEE 802.11ad: Directional 60 GHz communication for multi-Gigabit-per-second Wi-Fi [Invited Paper]. IEEE Communications Magazine. December 2014;**52**(12):132-141. DOI: 10.1109/MCOM.2014.6979964

[32] Verma L, Fakharzadeh M, Choi S. Wifi on steroids: 802.11AC and 802.11AD. IEEE Wireless Communications. December 2013;**20**(6):30-35. DOI: 10.1109/MWC.2013.6704471

[33] Biglarbegian B, Fakharzadeh M, Busuioc D, Nezhad-Ahmadi MR, Safavi-Naeini S. Optimized microstrip antenna arrays for emerging millimeter-wave wireless applications. IEEE Transactions on Antennas and Propagation. May 2011;**59**(5):1742-1747. DOI: 10.1109/TAP.2011.2123058

[34] Li M, Luk KM. Low-cost wideband microstrip antenna array for 60-GHz applications. IEEE Transactions on Antennas and Propagation. June 2014;**62**(6):3012-3018. DOI: 10.1109/TAP.2014.2311994

[35] Carles G, Oller J, Paradells J. Overview and evaluation of bluetooth low energy: An emerging low-power wireless technology. Sensors. 2012;**12**(9):11734-11753. DOI: 10.3390/s120911734

[36] IEEE standard for low-rate wireless networks. IEEE Std 802.15.4-2015 (Revision of IEEE Std 802.15.4-2011); April 22, 2016. pp. 1-709. DOI: 10.1109/IEEESTD.2016.7460875

[37] ZigBee Specification Version 3.0. December 2016. [online]. Available from: http://www.zigbee.org/zigbee-for-developers/zigbee/

[38] Montenegro G, Kushalnagar N, Hui J, Culler D. Transmission of IPv6 Packets Over IEEE 802.15.4 Networks. September 2007. [online]. Available from: http://www.ietf.org/rfc/rfc4944.txt

[39] Nolan KE, Guibene W, Kelly MY. An evaluation of low power wide area network technologies for the Internet of Things. In: 2016 International Wireless Communications and Mobile Computing Conference (IWCMC); Paphos; 2016. pp. 439-444. DOI: 10.1109/IWCMC.2016.7577098

[40] Margelis G, Piechocki R, Kaleshi D, Thomas P. Low throughput networks for the IoT: Lessons learned from industrial implementations. In: 2015 IEEE 2nd World Forum on Internet of Things (WF-IoT); Milan; 2015. pp. 181-186. DOI: 10.1109/WF-IoT.2015.7389049

[41] About SIGFOX. SigFox [online]. Available from: http://www.sigfox.com/en/#!/about

[42] Minh-Tien D, Claire G, Jean-Marie G. Interference modelling and analysis of random FDMA schemes in ultra narrowband networks. In: The Tenth Advanced International Conference on Telecommunications; AICT; 2014. pp. 132-137

[43] Roth Y, Doré J-B, Ros L, Berg V. Turbo-FSK, a physical layer for low-power wide-area networks: Analysis and optimization. Comptes Rendus Physique. Available online 13 December 2016, ISSN 1631-0705. DOI: http://dx.doi.org/10.1016/j.crhy.2016.11.005

[44] Roth Y, Doré JB, Ros L, Berg V. Turbo-FSK: A new uplink scheme for low power wide area networks. In: 2015 IEEE 16th International Workshop on Signal Processing Advances in Wireless Communications (SPAWC); Stockholm; 2015. pp. 81-85. DOI: 10.1109/SPAWC.2015.7227004

[45] LoRa Alliance [Internet].2016. Available from: www.lora-alliance.org. [Accessed: 2016-12-12]

[46] LoRa Alliance. LoRa Specification V1.0.2. Technical Report. July 2016

[47] Petajajarvi J, Mikhaylov K, Roivainen A, Hanninen T, Pettissalo M. On the coverage of LPWANs: Range evaluation and channel attenuation model for LoRa technology. In: 2015 14th International Conference on ITS Telecommunications (ITST); Copenhagen; 2015. pp. 55-59. DOI: 10.1109/ITST.2015.7377400

[48] Aloÿs A, et al. A study of LoRa: Long range & low power networks for the internet of things. Sensors. 2016;**16**(9):1466. DOI: 10.3390/s16091466

[49] Reynders B, Meert W, Pollin S. Range and coexistence analysis of long range unlicensed communication. In: 2016 23rd International Conference on Telecommunications (ICT); Thessaloniki; 2016. pp. 1-6. DOI: 10.1109/ICT.2016.7500415

[50] Shukla A, Karki H. Application of robotics in onshore oil and gas industry—A review Part I. Robotics and Autonomous Systems. January 2016;**75**(Part B):490-507. ISSN 0921-8890. DOI: http://dx.doi.org/10.1016/j.robot.2015.09.012

[51] Matthew L, Karingattil R,Williams L. ISM-band and short range device regulatory compliance overview [Internet]. Texas Instruments. 2005;(1):1-16. Available from: http://www.ti.com/lit/an/swra048/swra048.pdf. [Accessed: 2017-01-12]

[52] LINKPlanner. Cambium Networks. Cambium LINKPlanner User Guide 4.5.1. Available online at www.cambiumnetworks.com/products/software-tools/linkplanner/

[53] Semtech. SX1272/73 - 860 MHz to 1020 MHz low power long range transceiver. 2015. Available from: http://www.semtech.com/images/datasheet/sx1272.pdf [Accessed: 2016-09-07]

[54] Keshav S. Design principles for robust opportunistic communication. In: Proceedings of the 4th ACM Workshop on Networked Systems for Developing Regions (NSDR '10), Article 6; ACM, New York, NY, USA; 2010. p. 6. DOI: http://dx.doi.org/10.1145/1836001.1836007

Section 4

Legal Issues in Robotics

Chapter 8

Implications of the Google's US 8,996,429 B1 Patent in Cloud Robotics-Based Therapeutic Researches

Eduard Fosch Villaronga and Jordi Albo-Canals

Additional information is available at the end of the chapter

http://dx.doi.org/10.5772/intechopen.70279

Abstract

Intended for being informative to both legal and engineer communities, this chapter raises awareness on the implications of recent patents in the field of human-robot interaction (HRI) studies. Google patented the use of cloud robotics to create robot personality(-ies). The broad claims of the patent could hamper many HRI research projects in the field. One of the possible frustrated research lines is related to robotic therapies because the personalization of the robot accelerates the process of engagement, which is extremely beneficial for robotic cognitive therapies. This chapter presents, therefore, the scientific examination, description, and comparison of the Tufts University CEEO project "Data Analysis and Collection through Robotic Companions and LEGO® Engineering with Children on the Autism Spectrum project" and the US 8,996,429 B1 Patent on the Methods and Systems for Robot Personality Development of Google. Some remarks on ethical implications of the patent will close the chapter and open the discussion to both communities.

Keywords: cognitive therapeutic robots, cloud robotics, Google patent, personality, personalization, ASD research, ethical implications

1. Introduction

Compared to neurologically typical children, children and adolescents under the autistic spectrum disorder (ASD) have persistent deficits in social communication and social interaction across multiple contexts [1]. They normally have deficits in social-emotional reciprocity and difficulties in developing, maintaining and understanding relationships [2]. Helping these children to deal with a multitude of simultaneous sensory inputs and peer-mediated approaches through social play interventions has been proven to be effective [3]. The problem with traditional interventions, however, is that researchers are confronted with the task to investigate the complex relationship between the acquisition of communication skills, social-emotional factors and types of transactional support that predict better outcomes for children with ASD [4]. Moreover, this is greatly challenged by the fact that, albeit children with ASD have comparable developmental difficulties, there are many differences among children with ASD [5].

A therapy to be effective, therefore, not only should address the predominant core characteristics of ASD, but also be individualized to meet the needs of each participant [4]. Robots help bridge this existing gap because they can adapt easily to each individual's needs, they are predictive and repetitive and also very engaging [6, 7]. In fact, not only the use of robots has been found to be remarkable in cognitive rehabilitation therapies, but also the actual process of building them encourages social and cooperative skills, which can be very positive for autistic children [8–10].

The Tufts University CEEO project "Data Analysis and Collection through Robotic Companions and LEGO® Engineering with Children on the Autism Spectrum" measures the effect of LEGO® engineering and its collaborative nature on the development of social skills in children and adolescents with ASD. Furthermore, in order to contribute to solve the lack of quantitative data in projects concerning robots and autism [11], the project uses logger robots connected to a cloud system combined with a traditional recording and coding system to allow the data collection. The cloud system will also help control the behavior of the robots, which will participate actively in the classroom playing the role of master to help students work together and achieve classroom goals.

On March 31, 2015, Google was awarded a patent regarding methods and systems for robot personality development. The patent covers those robots that can be customizable with personality attributes and related capabilities drawn from cloud computing capacities [12]. According to the patent, these attributes can be in audio or visual format and can be derived from the human-robot interaction, the surroundings or the circumstances. Moreover, Google aims at the transferability of these robot personalities. For at least 20 years, Google will have the exclusive right to exploit the content of the patent.

The correlation between both the CEEO project and the Google patent lies on the fact that, in order to have success on the therapy, the children need to be engaged and this engagement comes, most of the times, from the robot personalization. Although the engagement between the user and the robot is not customization-dependent [13], it is found that the personalization of it accelerates the process of engagement. Indeed, because engagement drives learning, there are a lot of educational and therapeutic projects that personalize robots to promote this engagement [14, 15]. The trickiest part lies on the fact that this personalization is done through the cloud system, as it happens with the CEEO project.

Although patents cannot frustrate the primary object of the patent laws, i.e., to promote innovation [16], the broad method patented by Google could block all those projects that use robots and cloud services in the same line. This is the *raison d'être* of this article: by explaining the similarities between the CEEO project (Section 3) and the Google patent (Section 4), this book chapter will explain why the Google patent could frustrate ongoing projects in education—for neurotypical children and for nonneurotypical children [17, 18]. Basic concepts such as cloud, patent, robot system or personalization will introduce the discussion in Section II, as this chapter aims at being informative to both the legal and the technical communities. Some remarks on how this could be avoided will be shortly presented in Section 5 too.

2. Definitions

In order to fully understand the controversial situation about how Google's patent US 8,996,429 B1 can interfere with current and future research based on cloud robotic systems, we need to define the four key factors involved: what is cloud robotics, what is a robotic system, what does personality mean for robots and what are patents.

2.1. Cloud robotics

In 1997, Ibana was the first to give a premature explanation of cloud robotics. He explained, "a remote-brained robot does not bring its own brain with the body. It leaves the brain in the mother environment, by which we mean the environment in which the brain's software is developed, and talks with it by wireless links" [19].

This is based on what later on would be understood for *cloud computing*. Coined in 1996, and extended in 2006 [20], National Institute of Standards and Technology (NIST) defined cloud computing in 2011 as a "model for enabling ubiquitous, convenient, on-demand network access to a shared pool of configurable computing resources [...] that can be rapidly provisioned and released with minimal management effort or service provider interaction" [21].

Similar to the idea of Ibana, in 2010, Kuffner saw also the advantages of using cloud capacities for robots: it could provide a shared knowledge database, offload heavy computing tasks to the cloud and create a reusable library of skills or behaviors that map to perceived complex situations [22]. In the same year, others were also announcing cloud-computing frameworks for service robotics [23]. Soon, the concept of *cloud robotics* was consolidated, and nowadays, it refers to "any robot or automation system that relies on data or code from a network to support its operation, i.e., where not all sensing, computation and memory is integrated into a single standalone system" [24].

Even if a recent concept, most of roboticists having to process large quantities of data with their robot can choose to use cloud robotic platforms because all the constraints related to the existing framework whatsoever its nature (resources, information or communication) are somehow mitigated [25]. In fact, Inaba already conceived something similar called "remote-brained robot" [19], and even the Roboearth project introduced a World Wide Web for robots in 2011. In this project, it was argued that in the near future robots would need to reliably perform tasks beyond their explicitly preprogrammed behaviors and quickly adapt to the unstructured and variable nature of tasks, something unlikely without a cloud platform [26].

Waibel et al. proved that the use of a cloud system could create an environment where knowledge and information could be shared, allowing a better robot performance, and where robots could use this shared knowledge independently of their architecture. In addition, they expounded that it may also offer other benefits by allowing component reuse across different systems and developers, human knowledge about the component usage, robustness and efficiency. In the following years, several researches have adopted this technology to foster individual child partnership in medical facilities, for robot companions for elderly care or for education purposes [17, 27, 28].

2.2. Robot and robotic system

Although Čapek brothers' intention was not to define "robots," because they basically just ushered the word into existence [29], thanks to their 1922 play R.U.R. [30], Oxford dictionary today defines robots as "machines capable of carrying out complex series of actions automatically, especially one programmable by a computer." On its side, an although stating that the question was too meaningless to deserve discussion, Alan Turing in 1950 already believed that at the end of twentieth century people could talk about machines thinking without being contradicted [31].

Nowadays, and not far away from this vision, robots are considered machines, situated in the world, that sense, think and act [32]. Although it has been argued that a robot does not *think* but rather processes the information and weights potential outcomes [33], it is also true that the word "think" cannot be interpreted in its common meaning [34]. In fact, the machine decision-making process normally includes (1) data acquisition, perception through sensors (infrared, radar, stereovision, optical encoders, etc.) and filtering/fusing information; (2) navigation, localization and decision-making (path planning, obstacle avoidance and machine learning); and (3) locomotion, kinematics and motor control in order to act (in various forms: manipulating or moving) [35].

A robot therefore can sense its environment, has the capacity to process the information and is organized to act directly upon its environment [36]. *Mobility*, therefore, is an important aspect when defining robots. Consistently, the industry defines a robot as an "actuated mechanism programmable in two or more axes with a degree of autonomy moving within its environment, to perform intended tasks" [37].

A *robotic system* relates to all the systems that, in interaction with each other and the environment, allow the robot to actuate. In the case of the CEEO project, and at the physical level, the technology involved in the pilots is a robot companion for each group of learners, a wearable device per learner, a touchable device to interact with the robot, and a laptop to implement part of the sensors' signal processing. At the network level, the network electronics needed to provide the cloud services to run all the integrated system. At the application level, information about the performance of the child is stored. Cameras hidden in the room also provide information regarding the environment.

2.3. Personality of the robot

Personality refers to the "dynamic integration of the totality of a person's subjective experience and behavior patterns, including both (1) conscious, concrete, and habitual behaviors, experiences of self and of the surrounding world, conscious, explicit psychic thinking, and habitual desires and fears and (2) unconscious behavior patterns, experiences and views, and intentional states" [38]. UNESCO defines behavior as the way in which an individual behaves or acts, even if there is not an accepted definition of behavior [39, 40]. In plain language, personality is what an individual is, and behavior is what an individual does. What do personality and behavior mean with regard to robots?

Industrial robots did not have interaction with humans—they were normally fenced off to protect humans. The concept of personality and behavior of robots began with the inception of social robots. Already in 1999, Breazeal and Scassellatti were working on robots that could interact socially with humans [41]. Miwa et al. highlighted in 2001 that the personality of the robot was especially important in achieving smooth and effective communication with humans [42]. In 2003, Fong et al. presented a review of the common features of social robots [43]. According to them, social robots expressed/perceived emotions, communicated in high-level dialogue, learned/recognized models of other agents, established/maintained social relationships, used natural cues—such as gaze or gestures—and exhibited distinctive personality and character and that might learn/develop social competencies. They agreed that social robots could be very different, since those robots uniquely engage people in social interactions, to robots that were programmed to fulfill social norms and carry out tasks in environments habited by humans. They also mentioned that some of these robots use deep models of human interaction to proactively encourage social interaction, while others would rely on humans to attribute mental states and emotions to the robot. To this, and similar to the idea that the complexity of the behavior of an ant is more a reflection of the complexity of its environment than its own internal complexity (speculated that the same may be true for humans) [44], the environment can influence the behavior of a robot directly, through sensors, or indirectly by the action of the user.

In 2006, a large study on the personality of social interactive robots and human perceptions was conducted [45]. The participants of the study perceived the robot's personality although its nature was nonhuman. This has had an impact on human-robot interaction studies as the personalization of the robot—meaning adapting its personality to the user—is widely accepted to play a major role in accelerating the engagement with the robot, partly because it motivates the user [13, 15].

2.4. Patents

The United States was the last country to adopt a first-to-invent patent system. In 2011, however, following other examples like EP, JP and CN, the United State Congress passed the Leahy-Smith America Invents Act (AIA), which involved the abolishment of the long tradition first-to-invent system and the adoption of the first-to-file system [46]. After the AIA entered into force in 2013, the Office Patent would grant the patent to whoever filed the application first regardless of its invention date (with some exceptions and grant periods previewed in 35 USC §102) [47].

§101 of the above-mentioned *corpus iuris* expounds that "whoever invents or discovers any new and useful process, machine, manufacture, or composition of matter, or any new and useful improvement thereof, may obtain a patent [...]." In other words, a patent is a "document, issued, upon application, by a government office [...] which describes an invention and creates a legal situation in which the patented invention can normally be only exploited [...] with the authorization of the owner of the patent" [48].

Not all inventions can be patented though. Actually, inventions need to meet some general conditions to be patentable. An invention will be patentable if it:

- has a patentable subject matter (i.e., it falls under the definition of 35 USC §101),
- is novel (i.e., it is something new according to 35 USC §102), and
- is nonobvious (i.e., nonobvious for those who have an ordinary skill in the art, 35 USC §103).

Once granted, the patent holder will have a period of time to exploit it exclusively. Article 47 of the Patent Cooperation Treaty leaves the contracting parties to decide time limits. In United States, time limit for patent exploitation is at least 20 years, unless failure on the payment of the maintenance fee (vid. 35 USC §41.c.1).

3. Robotic therapies: the case of the robot companions and LEGO® engineering

LEGO®-based robots have been proved to be an effective tool to use in education, not only for undergraduates, but actually for all ages [49, 50]. LEGO engineering creates a context where social and problem-solving skills meet each other. This has been found extremely positive for children with neurodevelopmental disorders [51–53]. It seems that making the robot behavior depend on user actions is positive in robot therapies because it involves the motivation of the user [54].

The CEEO project aims at observing and measuring the engineering skills and processes of children with ASD using this type of robots. The project aims at collecting quantitative data to compare the results with those of typically developing children. The main idea is to look for examples where students with ASD can be role models for typically developing students, both in how they develop and in how they solve engineering problems.

Implementing this methodology with high dysfunctional children, nevertheless, would require a lower ration of children than therapists, needless to say that the collection of information from the session would be very hard to obtain. To solve these issues, the CEEO project introduces a cloud-based robotic system that includes (**Figure 1**): social robots that can work as mediators, companions, which are connected online with an expert system developed from previous experiences and human experts in the field of ASD, as well as external sensors such as cameras or user interfaces. This way two of the main tenets of practice in intervention approaches for autistic children are met [4]: individualization of the therapy to match children's current developmental level on his or her profile, because the exercises are done in accordance with the child's performance level, and the address of the predominant core characteristics of ASD, because all the exercises are focused on social-emotional reciprocity, verbal communication and cognitive processing similar to previous studies of the same researchers [55].

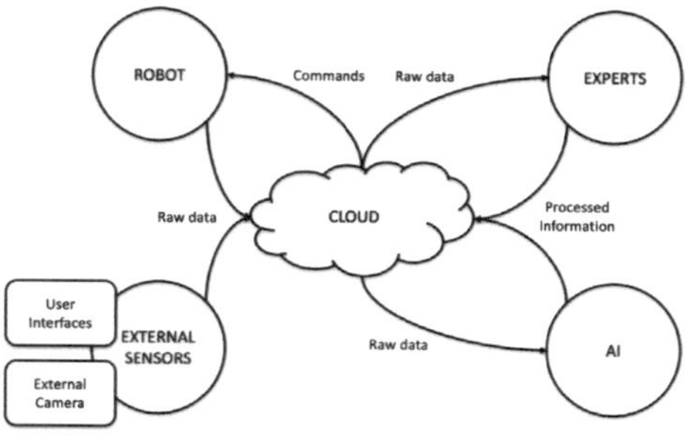

Figure 1. Robot and Cloud Interaction Tufts-CEEO's project.

The robot companion has the role of a helper, a social mediator and a facilitator and reminds children of the time schedule. They interact with the children through gestures and expressions, lights, sounds and speech. The robot companions are controlled by the cloud system. Decisions are based on (1) the interface used by the conductor of the session/teacher (highest priority), (2) the web-based interface the learners use to communicate with the robot and (3) a probabilistic decision model based on past events and finite-state machines. The transition between states is produced by time schedule, as well as from the input from the children, or the teacher/instructor of the class.

Through the expertise acquired by the cloud system, it is expected that: (1) the robot can identify stressful situations and act as a companion to help cope and provide tailored strategies to the individual child throughout the social skills treatment plan; (2) the robot can work as a data logger that collects quantitative data, to understand how children with ASD deal with social situations and what strategies they use to solve problems; and (3) the complexity of the therapy is reduced in cost and time terms.

The cloud architecture helps maximize the effectiveness of the shared resources of the robotic system that is connected to a network [56]. In particular, the cloud-based infrastructure allows:

- The robots to upload video during the sessions. The video is stored on the server database. There is a camera on the ceiling of the room that is also sending the video recordings to the server database.
- The web-based child-to-robot interface to send commands to the cloud. The child uses it to interact with the robot companion.
- The web-based teacher/conductor interface to send commands to the cloud. The teacher uses to interact with the robot companion and also sends commands to the cloud.
- The robot companion to receive commands from the cloud that teleoperate their behavior.
- The robot companion to behave upon the information of the experts on ASD and the therapists—to detect stressful situations for instance.
- Researchers to login to the cloud system to watch and code the video in order to provide human feedback to the artificial intelligent (AI) system.

- The information from the questionnaires to be stored in the server database and to be seen by the researchers.
- The process of the acquired information to create models and descriptors of the interaction state of the children.

Robots in the project have a preprogrammed personality because it is expected to interact with the children on a high-level dialogue and use natural and social cues [43]. For emotion recognition exercises, the companion robot is expected to express emotions so that the child can perceive them as such. The robot behaves according to this by-default personality.

There is evidence in the literature that a good match between a patient and a coach produces better results of therapy or treatment [57]. And because algorithms that learn how to customize certain objects to customize personal characteristics have existed now for more than 10 years [58], the robot, over time, changes its behavior. The personalization of the robot consists on building loyalty between the robot and the child through matching each other's needs, through the construction of a meaningful one-to-one relationship.

In the current project, the personalization of the robot is based on information the researchers get from the parents (through questionnaires). Sometimes the robot's personality changes according to the child's likes and dislikes. This information is drawn from the cloud. Only by understanding the needs of each individual, and by satisfying a goal that efficiently and knowledgeably addresses each individual's need in a given context, this can happen.

4. Google's patent: methods and systems for robot personality development

Google was granted a patent for methods and systems for robot personality development. This disclosure patented a process to create robots permeated with personality or personalities, drawn from cloud computing capacities, and capable of interact with users [12].

The scope of Google's patent is to describe the techniques and processes for user-robot interaction (URI) to engender personality for the robot. Google defines personality as the "personification in the sense of human characteristics or qualities attributed to a non-human thing [...] such that the robot interface is customized to provide a desired personality for the robot" [12].

The robot collects mass information from the user and its surroundings and tailors a personality to interact upon with the user. The ultimate goal is to interact with the user more personally. The patent describes what is the method that will use to collect the information to forge the personality of the robot, i.e., from different sources and by the processing of all this information in the cloud. Some examples of raw data and devices to which the robot might have access are briefly enounced in the patent:

- The robot might have information from the user himself/herself, e.g., all the possible information relating to: calendar, email, text messages (or other electronic correspondence), call log, recently accessed documents on a computer, Internet browser history and so on;
- The robot might have access to the user's devices, which could include: a computer, laptop, mobile phone, PDA, tablet, cellular or other mobile computing devices. Any other television or cloud computing devices, or any device with capacity to access the cloud, will also be considered as a user's device;
- The robot's sensors could collect information about the environment such as: the location, time of the day or weather, and even information about nearby objects, the language the user uses or information that can be available through the interaction with other robots;
- The robot might have access to the information stored in online profiles the user might have on the Internet, e.g., social network sites.

The robot might send all this unstructured data to the cloud and receive back processed data to customize the personality of the robot. As we can see in **Figure 2**, the interaction could not only be between the user and the robot, but also between the robot and other sources, such as other sensors, other robots or the Internet itself. These latter interactions could be done directly, e.g., between robots, or indirectly, through the cloud. Indeed, it will be able to share information with other cloud computing devices.

Google's patent includes the robot's estimation of the user's mood. The idea is to evoke positive responses when the user feels sad, either computationally or locally if a mood recognition database has been provided to the robot [12]. In this model, moreover, Google envisages the possibility of transferring the robot personality, through the cloud, to other robots.

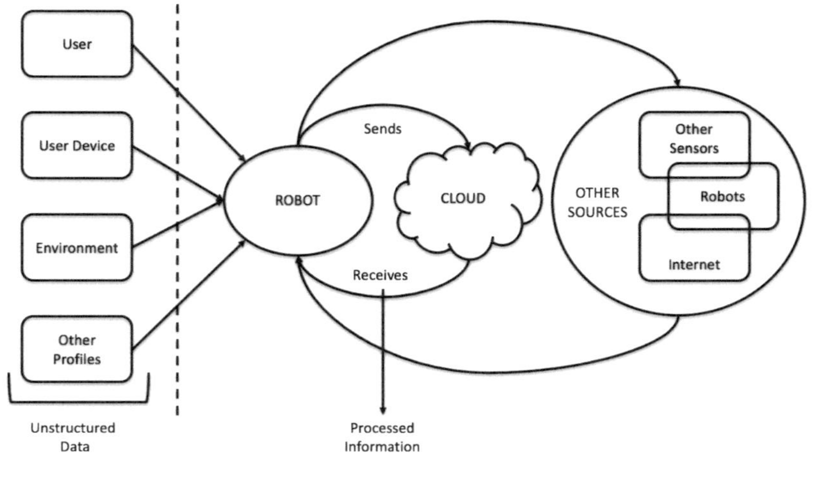

Figure 2. Robot and Cloud Interaction. Google's US 8,996,429 B1 Patent.

5. Adequacy of the patent to patent law

US 8,996,429 B1 Google Patent is called into question below. Using the Alice US Supreme Court judgment as a base, the patent is questioned about its adequacy to the conditions for patentability established in title 35 United States Code [59]. As established in the case Verdegaal Bros v. Union Oil Co. of California, we will try to argue if each and every element of the claim has been addressed previously, because a claim is anticipated only if each element is found, either expressly or inherently described, in a single prior art reference [60].

5.1. Patentable subject matter

The definition set in 35 USC §101 suggests a very important exception: laws of nature, natural phenomena and abstract ideas are not patentable [16]. Patents are the basic tools for scientific and technological work, and they tend to promote innovation, not to impede it, and that is why these aspects are not patentable [61]. Google, nonetheless, patented a broad method—the creation of a robot personality using cloud robotics—that might impede innovation.

The Supreme Court, however, states that an invention is not directly invalid because it is an abstract concept, but it will remain eligible for a patent when those concepts can be applied to a new and useful end [62]. As described in the next subsection, Google patents a well-known method to process all the collected information for the creation of the personality of the robot, which is not something new: all social robots have a default personality [43]. Breazeal stated that the physical appearance, the robot manners of movement and its manner of expression convey personality traits to the person who interacts with [63]. According to her, this fundamentally influences the manner in which people engage the robot. From 1997 to 2000, they already developed Kismet with (infant-level) social competences that were already running in fifteen computers.

The Court adds in the Mayo judgment that, in order to be eligible for a patent, the application of the law-of-nature, natural-phenomenon and abstract-idea concepts must be determined to be an inventive concept [61]. Detecting the user's mood through sensors, nevertheless, and using cloud-computing capabilities to process all the information and modify thereupon the personality of the robot is not an inventive concept [27, 64, 65].

5.2. New (novel)

35 USC §102 describes generally speaking that only new inventions can be patented. Google describes the idea of using cloud-computing capabilities to collect mass data, reduce/offload the intensive workloads from the onboard resources on robots, to create a robot personality and to transfer this personality from one robot to another one. All these procedures/methods/concepts have been done before:

- Regarding workload reduction in robots, Ibana already envisaged the idea of a "remote-brained robot" although the term "cloud computing" was not yet used in 1993 [19]. Softbank has marketed the use of cloud AI and an emotion engine for a robot already as a product [66].

- The idea of creating a robot that could interact with information from the physical world dates from 1993 too [67]. Brooks and Stein wanted to design a robot that could "learn new behaviors under human feedback such as human manual guidance and vocal approval."
- The use of sensors to collect grounded and real information from the user is not new [27].
- Transferring the collected information to a single collection point, the possibility to share it among robots (robots talking to their neighbors) is not new. Winfield addressed largely the way collective robots work [68].
- Building a personality in a robot is not new either. Breazeal and Fong et al. already stated in the early 2000 that socially interactive robots exhibit distinctive personality. Studies regarding robot personality and user's perception have been carried out during 2006 [45].
- In 2008, moreover, Wowwee® released Mr. Personality™, a robot that had personality. His user manual states that Mr Personality™ comes with a default persona (Max and Simon), similar as what Google describes on its patent, and the user can use software to download new personalities via Internet [69]. Multiple personalities, therefore, are not a new concept either.
- Transferrable personalities are neither a new concept. Page 35 of the user manual of the Wowwee® robot states that: "to transfer personalities from your computer to your robot click on MyComputer [...]."

Both the method and the hardware have been already addressed in previous literature and products. Furthermore, the construction of personality for robots to interact with humans is one of the very basic foundations of human-robot interaction studies.

5.3. Be nonobvious (inventive step)

35 USC §103 strictly says "a patent may not be obtained [...] if the differences between the subject matter sought to be patented and the prior art are such that the subject matter as a whole would have been obvious at the time the invention was made to a person having ordinary skill in the art to which said subject matter pertains [...]."

The problem is that cloud robotics is a logical step toward solving the problems networked robots are encountering [26]. Indeed, the use of cloud robotics mitigates mostly all the constraints related to their existing framework [25]. The way Google expounds it reminds to what happened in Bilski v. Kappos [70], as if the whole method was "taught in any introductory [in this case computer science] class."

Actually, Google patents some well-known procedures to collect data from different sources, to use this data to personalize the robot [58]. The jurisprudence of US Supreme Court is clear in this respect: "simply appending conventional steps, specified at a high level of generality, to a method already 'well known in the art' is not 'enough' to supply the 'inventive concept' needed to make this transformation" [59].

Indeed, similar to what the Alice judgment stated, "the functions performed by the computer at each step of the process are purely conventional and considered as an ordered combination the computer components of petitioner's method add nothing that is not already present when the steps are considered separately" [59], one could understand that if instead of *computer* the wording read *robot* then the Court would be referring to the case we are addressing.

6. Other remarks: ethical questions arisen from the patent

The Google patent describes what is the method to create the personality of the robot and uses examples to represent its possibilities. These examples raise several ethical questions. Here we will mention: (1) the safety of the user (produced by reinforcing learning); (2) the possibility to talk to dead people; and (3) the delegation of autonomy (which could lead to robot responsible scenarios).

1. The patent of Google states:

   At block 826, the method 820 includes modifying the default user-profile to incorporate the estimated personality so as to provide a modified persona. This can be a transitory modification or something more permanent. For example, the robot may prepare food for the user using peanut oil. The user, who may be allergic to peanut-based foods, may eat the meal and have an allergic reaction. The user may further scold the robot for cooking the meal with peanut oil. Scolding may be considered a negative feedback response where the user is directing a negative reaction toward the robot for an action that the robot committed. On the other hand, a positive feedback response may be a positive reaction toward the robot for an action that the robot committed. In this example above, the robot may permanently modify information in the user-profile to include the user's allergic reaction to peanut and avoid anything to do with peanuts in the future.

In this paragraph, the patent argues that the robot can estimate the user's mood depending on the reactions to a certain scenarios. This estimation nevertheless can lead to a serious critical risk scenario. Food-induced anaphylaxis affects multiple organ systems and hospitalization due to the fact that it has increased over these years [71, 72]. Even if the patent refers to a particular scenario, and wants to emphasize the fact that depending on the reactions the robot will be able to discover whether the user might like one thing or another thing (as a kind of reinforce learning), this constitutes an overtaking decision-making process from the robot. Independently of whether a robot can or cannot learn from the experience of the user, the actions autonomously taken by the robot should never endanger the safety of the user. Even if the robot might be in a learning process, there are several protective measures that should be embedded to avoid any unfortunate scenario. If the robot prepares the meal with peanut oil, the person suffers anaphylaxis, and then, the system fails to call an ambulance; then, not only the company would be responsible for an unwilling scenario, it will be responsible for the death of a person.

As suggested by Amodei et al., "systems that simply output a recommendation to human users, such as speech systems, typically have relatively limited potential to cause harm. By contrast, systems that exert direct control over the world, such as machines controlling industrial processes, can cause harms in a way that humans cannot necessarily correct or oversee" [73]. This should be carefully addressed, especially in the light of what the patent describes "the robot may respond to the negative reinforcement response by continuing to perform other tasks until a positive reinforcement response is received."

2. As written on the patent, "the robot may be programmed to take on the personality of real-world people […] or a deceased loved one." The patent suggests that with their method, and because there will be no deletion of the data, there will be the possibility to speak with the personality created from a person that could be dead or alive. This is the first time Google mentions death. Up to now, no provision regarding death can be found in its terms and conditions [74]. Postmortem privacy has been addressed by other platforms like Facebook [75]. In Europe, the 679/2016 General Data Protection Regulation does not address this topic, even if there are some EU member states that have decided to cope with it [76]. The Article 29 Working Party said in an opinion, "information relating to dead individuals is therefore not to be considered as personal data" even if "may still indirectly receive some protection" [77]. In the light of the intentions of the Google patent, and lacking an express provision in this regard, it will be extremely important to answer the question whether this function of the (possible future) robot of Google is ethical or not, and how this should be modeled.

3. To close, Google aims the robot to take over in several situations. As an example the patent states, "the robot may then adopt a persona of the user's mother, and indicate 'it is time to clean out the refrigerator, honey'" [12]. Delegation of authority in the human decision-making process—to a robot in this case—nevertheless, calls for special attention. In sociology it is said that one actor has authority over another when the first holds the right to direct the actions of the second [78]. Linked to it, if robots have agenthood (an hypernym to describe that not only humans exhibit morally responsible behavior) [79]) as the European Parliament suggests on its latest resolution [80], then it could possibly mean that the robot is held responsible for its acts—which may lead to held Google responsible for them as it occurs with the autonomous cars.

To all this, there are currently no legally binding frameworks or guidelines on the creation of robotic technology that could approach ethical implications. The only corpus addressing this issue is "BS 8611: Robots and robotic devices—Guide to the ethical design and application of robots and robotic systems," which was recently published. BS 8611 has identified broad range of ethical hazards and their mitigation including societal, application, commercial/financial and environmental risks. Concerning societal hazards, the concepts of deception, privacy, confidentiality, addiction, loss of trust and employment are also addressed.

However, it is uncertain to what extent social robot creators should incorporate features beyond mere technical aspects in their design process—as the BS 8611:2016 standard suggests. Standards are good instruments to deal with complex, new and international issues; however, they do not have binding force. The draft report of the European Parliament is a *lege ferenda* (a proposal for a future law) and has been approved by the European Commission in January 2017. If this ends up into a binding corpus iuris, then an ethical framework for those who design robots will have to apply regardless of any patent.

7. Conclusion

Google secures the patent adding the following statement: "it should be understood that arrangements described herein are for purposes of example only. As such those skilled in art will appreciate that other arrangements and other elements [...] can be used instead." This basically means that no matter what machines or functions are used, if they are used for all the purposes described in the patent, they will under the patent's scope. This does not promote innovation according to the Mayo judgment, especially knowing that Google is not actually developing the technology but focusing on industrial robots, as Darling already mentioned in [81].

Many research projects, which could prove beneficial to children with autism, could be hampered by restrictions due to the broad claims of the Google patent. There exists, however, various administrative and judicial avenues to review the scope and validity of the patent claims. In September 2012, AIA incorporated the Inter Partes Review (IPR). The IPR is a procedure to challenge the validity of patent claims based on patent and on printed publications. As stated by the PLI Patent Bar Review, the IPR proceedings are available for any patent whether issued before, on or after September 16, 2012. The transitional program for covered business method patents applies to any covered business patent in the same terms. Furthermore, if there was an action, the Supreme Court of US could decide whether the patent was correctly granted or not.

Author details

Eduard Fosch Villaronga[1]* and Jordi Albo-Canals[2]

*Address all correspondence to: e.foschvillaronga@utwente.nl

1 University of Twente, Enschede, The Netherlands

2 NTT, Tufts University, USA

References

[1] American Psychiatric Association. Diagnostic and Statistical Manual of Mental Disorders, 5th ed., 2013, Washington, DC

[2] Sicile-Kira C. Autism Spectrum Disorder (Revised): The Complete Guide to Understanding Autism. Tarcher Perigee; Penguin Books, United States, 2014

[3] White SW, Keonig K, Scahill L. Social skills development in children with autism spectrum disorders: A review of the intervention research. Journal of Autism and Developmental Disorders. 2007;**37**(10):1858-1868

[4] Prizant BM, Rubin E. Contemporary issues in interventions for autism spectrum disorders: A commentary. Journal of the Association for Persons with Severe Handicaps. 1999;**24**(3):199-208

[5] Greenspan SI, Wieder S. Developmental patterns and outcomes in infants and children with disorders in relating and communicating: A chart review of 200 cases of children with autistic spectrum diagnoses. Journal of Developmental and Learning Disorders. 1997;**1**:87-142

[6] Barco A, et al. A drop-out rate in a long-term cognitive rehabilitation program through robotics aimed at children with TBI. In: The 23rd IEEE International Symposium on Robot and Human Interactive Communication, 2014 RO-MAN, 25-29th August, Edinburgh, United Kingdom. IEEE; 2014. pp. 186-192

[7] Valenzuela E, et al. Learning social skills through LEGO-based social robots for children with autism spectrum disorder at CASPAN center in Panama. In: Heerink M, de Jong M, editors. Conference Proceedings of New Friends 2015: The 1st International Conference on Social Robots in Therapy and Education; Windesheim Flevoland, 22-23 October, Almere, the Netherlands, 2015

[8] Michaud F, et al. Perspectives on mobile robots as tools for child development and pediatric rehabilitation. Assistive Technology: The Official Journal of RESNA. 2007;**19**(1):21-36

[9] Caci B, et al. New frontiers for psychology and education: Robotics. Psychological Reports. 2004;**94**(3 Suppl):1372-1374

[10] Dautenhahn K. Design issues on interactive environments for children with autism. In: Proceedings of ICDVRAT 2000: The 3rd International Conference on Disability, Virtual Reality and Associated Technologies. University of Reading; 23-25th September, Alghero, Italy, 2000

[11] Scassellati B, et al. Robots for use in autism research. Annual Review of Biomedical Engineering. 2012;**14**:275-294

[12] A. G. Francis and Jr., T. Lewis, Methods And Systems For Robot Personality Development: Nonlinear Resonant Circuit Devices. Google. U.S. Patent No. 8 996 429 B1, March 31, 2015

[13] Sung JY, et al. My Roomba is Rambo: Intimate home appliances. In: International Conference on Ubiquitous Computing. 16-19th September, Inssbruck, Austria, 2007. Springer Berlin, Heidelberg: Springer; 2007. pp. 145-162

[14] Christenson SL, et al., editors. Handbook of Research on Student Engagement. Springer Science & Business Media; Springer-Verlag New York , United States, 2012

[15] Andrist S, et al. Look like me: Matching robot personality via gaze to increase motivation. In: Proceedings of the 33rd Annual ACM Conference on Human Factors in Computing Systems. 18-23rd April, Seoul, Korea, ACM; 2015. pp. 3603-3612

[16] Mayo Collaborative Services v. Prometheus Laboratories, Inc. 566, U.S., 2012

[17] Park IW, Han J. Teachers' views on the use of robots and cloud services in education for sustainable development. Cluster Computing. 2016;**19**(2):987-999

[18] EU H2020 Dream Project. Cfr.: www.dream2020.eu and www.utwente.nl/ctit/research/research_ projects/international/horizon2020%20-%20IA's/de-enigma.html

[19] Ibana M. Remote-brained robots. In: Proceedings of the 15th International Joint Conference on Artificial Intelligence (IJCAI-97). August 23-29, NAGOYA, Aichi, Japan, 1997. pp. 1593-1606

[20] Regalado A. Who coined 'Cloud Computing'? MIT Technology Review. 2011

[21] Mell P, Grance T. The NIST definition of cloud computing. Special Publication 800-145. September 2011. p. 2

[22] Kuffner J. Cloud-enabled humanoid robots. In: Humanoids 2010 Workshop "What's Next". Google Research. The Robotics Institute, Carnegie Mellon University; Pittsburgh, United States, 2010

[23] Arumugam R, et al. DAvinCi: A cloud computing framework for service robots. In: IEEE International Conference on Robotics and Automation (ICRA). May 3 – 7, Anchorage, Alaska, 2010. pp. 3084-3089

[24] Kehoe B, et al. A survey of research on cloud robotics and automation. IEEE Transactions on Automation Science and Engineering. 2015;**12**(2):398-409

[25] Hu G, et al. Cloud robotics: Architecture, challenges and applications. IEEE Network. 2012;**26**(3):21-28

[26] Waibel M, et al. A World Wide Web for robots. IEEE Robotics & Automation Magazine. 2011

[27] Navarro J, et al. A cloud robotics architecture to foster individual child partnership in medical facilities. In: Cloud Robotics Workshop in 26th IEEE/RSJ International Conference on Intelligent Robots and Systems. 3-7 Nov. Tokyo Japan, 2013

[28] Rodić A, et al. Development of human-centered social robot with embedded personality for elderly care. In: Wenger, P., Chevallereau, C., Pisla, D., Bleuler, H., Rodić, A. (Eds.) New Trends in Medical and Service Robots. Springer International Publishing; 2016. pp. 233-247

[29] Jones R. Personhood and Social Robotics—A Psychological Consideration. Routledge; London, United Kingdom, 2015. p. 53. Some years after the inclusion of the word "robot" into the Oxford Dictionary, K. Čapek wrote a letter to the dictionary Editor stating the fact that he was not the inventor of the word but his brother Josef Čapek, see: blog.oxforddictionaries.com/2012/11/thanksgiving/

[30] Čapek K. R.U.R. (Rossum's Universal Robots). Penguin, New York, United States, 2004

[31] Turing AM. Computing machinery and intelligence. Mind. 1950;**59**(236):433-460

[32] Bekey G. Current trends in robotics: Technology and ethics. In: P. Lin, K. Abney and G. A. Bekey. Robot Ethics: The Ethical and Social Implications of Robotics. Cambridge: MIT Press; 2012. pp. 17-34

[33] Lutz C, Tamò A. RoboCode-ethicists: Privacy-friendly robots, an ethical responsibility of engineers? In: Proceedings of the 2015 ACM SIGCOMM Workshop on Ethics in Networked Systems Research. 21 August, London, United Kingdom, 2015. pp. 27-28

[34] Turing AM. Can a machine think? The World of Mathematics. 1956;**4**:2099-2123

[35] Klier J. LabVIEW Robotics: Sense-Think-Act—Application Engineers National Instruments at the NIDays Conference. 2010. Available from: www.slideshare.net/mayankagarwal51/labview-robotics-sensethinkact last access 21 July 2017

[36] Calo R. Robotics and the lessons of cyberlaw. California Law Review. 2015;**103**:513-564

[37] ISO 8373:2012 Robot and Robotics Devices—Vocabulary

[38] Kernberg OF. What is personality? Journal of Personality Disorders. 2016;**30**(2):145-156

[39] W. Guez and J. Allen, Regional Training Seminar on Guidance and Counselling. Module 4: Behaviour Modi- fication. UNESCO; 2000

[40] Bergner RM. What is behavior? And so what? New Ideas in Psychology. 2011;**29**:147-155

[41] Breazeal C, Scassellati B. How to build robots that make friends and influence people. In: Proceedings of 1999 IEEE/RSJ International Conference on Intelligent Robots and Systems, 17-21 October, Kyongju, Korea 1999. IROS'99. Vol. 2. IEEE; 1999. pp. 858-863

[42] Miwa H, et al. Experimental study on robot personality for humanoid head robot. In: Proceedings of 2001 IEEE/RSJ International Conference on Intelligent Robots and Systems, 29 Oct – 3 Nov, Maui, United States, 2001. Vol. 2. IEEE; 2001. pp. 1183-1188

[43] Fong T, et al. A survey of socially interactive robots. Robotics and Autonomous Systems. 2003;**42**:145

[44] Simon HA. The Science of the Artificial. Cambridge, Massachusetts: MIT Press; 1996

[45] Lee KM, Peng W, Jin SA, Yan C. Can robots manifest personality?: An empirical test of personality recognition, social responses, and social presence in human–robot interaction. Journal of Communication. 2006;**56**(4):754-772

[46] Public Law 112-29. Leahy-Smith America Invents Act. An act to amend title 35, United States Code, to provide for patent reform. September 2011

[47] United States Code, Title 35—Patents

[48] WIPO Publication 489 (E). Intellectual Property Handbook: Policy, Law and Use. World Intellectual Property Organization (WIPO); Bern, Switzerland, 2004, p. 17

[49] Danahy E, et al. LEGO-based robotics in higher education: 15 years of student creativity. International Journal of Advanced Robotic Systems. 2014;**11**(2):27

[50] Mataric MJ. Robotics education for all ages. In: Proceedings of the AAAI Spring Symposium on Accessible Hands-on AI and Robotics Education. 22-24 March, Palo Alto, United States, 2004

[51] Dautenhahn K, Werry I. Towards interactive robots in autism therapy: Background, motivation and challenges. Pragmatics & Cognition. 2004;**12**(1):1-35

[52] Wainer J, et al. The effectiveness of using a robotics class to foster collaboration among groups of children with autism in an exploratory study. Personal and Ubiquitous Computing. 2010;**14**(5):445-455

[53] Albo-Canals J, et al. Comparing two LEGO robotics-based interventions for social skills training with children with ASD. In: RO-MAN, The 22nd International Symposium on Robot and Human Interactive Communication, 26-29th August, Gyeongju, Korea, IEEE, 2013

[54] Robins B, et al. Developing scenarios for robot assisted play. In: Proceedings of the 17th IEEE International Symposium on Robot and Human Interactive Communication, RO-MAN, 1-3 August, Munich, Germany, 2008

[55] Diaz M, et al. Robot assisted play with a mobile robot in a training group of children with autism. IEEE/RSJ International Conference on Intelligent Robots and Systems October 7-12, 2012, Vilamoura, Algarve Portugal IROS 2012

[56] Albo-Canals J, et al. Enhancing long-term children to robot interaction engagement through cloud connectivity. In: Proceedings of the Tenth Annual ACM/IEEE International Conference on Human-Robot Interaction, Extended Abstracts, 2-5 March, Portland, United States, ACM; 2015. pp. 105-106

[57] Tapus A, et al. User—robot personality matching and assistive robot behavior adaptation for post-stroke rehabilitation therapy. Intelligent Service Robotics. 2008;**1**(2):169-183

[58] Riemer DK, Totz C. The many faces of personalization. In: M. M. Tseng, F. T. Piller. The Customer Centric Enterprise. Berlin, Heidelberg: Springer; 2003. pp. 35-50

[59] Alice Corporation PTY. LTD. v. CLS Bank International, et al. 13-298, U.S., 2014

[60] Verdegaal Bros v. Union Oil Co. of California. 814 F.2d 628, 631 (Federal Circuit, 1987)

[61] Association for Molecular Pathology v. Myriad Genetics, Inc. 569, U.S., 2013

[62] Diamond v. Diehr. 450 U.S., 175, 187. 1981

[63] Breazeal CL. Sociable machines: Expressive social exchange between humans and robots [Doctoral dissertation]. Massachusetts Institute of Technology; Cambridge, United States, 2000

[64] Chella A, et al. An architecture for humanoid robot expressing emotions and personality. In: Proceedings Frontiers in Artificial Intelligence and Applications. Conference: Biologically Inspired Cognitive Architectures 2010 - Proceedings of the First Annual Meeting of the BICA Society, Washington, DC, USA, November 13-14, 2010

[65] Merrick KE. A comparative study of value systems for self-motivated exploration and learning by robots. IEEE Transactions on Autonomous Mental Development. 2010;**2**(2):119-131

[66] Pepper Robot, SoftBank Corp. Cfr.: https://www.ituaj.jp/wp-content/uploads/2015/10/nb27-4_web_06_ROBOTS_pepper.pdf

[67] Brooks RA, Stein LA. Building Brains for Bodies, AI Memo No. 1439. Massachusetts Institute for Technology; Cambridge, United States, 1993

[68] Winfield A. Distributed sensing and data collection via broken ad hoc wireless connected network of mobile robots. Distributed Autonomous Robotic Systems. 2000;**4**:273-282

[69] Mr. Personality Robot Manual. Cfr.: http://www.theoldrobots.com/images7b/Mr._Personality_Manual.pdf

[70] Bilski v. Kappos. 561, US 593, 2010

[71] Hogan SP, et al. Food-induced anaphylaxis: Mast cells as modulators of anaphylactic severity. Seminars in Immunopathology. 2012;**34**(5):643-653

[72] Turner PJ, et al. Increase in anaphylaxis-related hospitalizations but no increase in fatalities: An analysis of United Kingdom National Anaphylaxis Data, 1992-2012. Journal of Allergy and Clinical Immunology. 2015;**135**(4):956-963

[73] Amodei D, Olah C, Steinhardt J, Christiano P, Schulman J, Mané D. Concrete problems in AI safety. arXiv preprint arXiv:1606.06565. 2016

[74] Google Policy Terms. Cfr.: www.google.com/intl/en/policies/terms/

[75] McCallig D. Facebook after death: An evolving policy in a social network. International Journal of Law and Information Technology. 2014;**22**(2):107-140

[76] Maximillian Schrems v. Data Protection Commissioner, Case C-362/14, Cfr. http://curia.europa.eu/juris/document/document.jsf?text=&docid=169195&pageIndex=0&doclang=EN&mode=lst&dir=&occ=first&part=1&cid=89046

[77] Article 29 Data Protection Working Party. Opinion 04/2007 'On the Concept of Personal Data'. June 2007. pp. 22-23

[78] Coleman JS. Foundations of Social Theory. Harvard University Press; Cambridge, United States, 1994. p. 66

[79] Floridi L, Sanders JW. On the morality of artificial agents. Minds and Machines. 2004;**14**(3):349-379

[80] 2015/2103(INL). European Parliament Draft Report on Civil Law Rules on Robotics. February 2017

[81] Darling K. Why Google's robot personality is not good for robotics. IEEE Spectrum. April 2015